T0235206

Lecture Notes in Artificial Intelligence   10004

Subseries of Lecture Notes in Computer Science

LNAI Series Editors

Randy Goebel
  University of Alberta, Edmonton, Canada
Yuzuru Tanaka
  Hokkaido University, Sapporo, Japan
Wolfgang Wahlster
  DFKI and Saarland University, Saarbrücken, Germany

LNAI Founding Series Editor

Joerg Siekmann
  DFKI and Saarland University, Saarbrücken, Germany

More information about this series at http://www.springer.com/series/1244

Masayuki Numao · Thanaruk Theeramunkong
Thepchai Supnithi · Mahasak Ketcham
Narit Hnoohom · Patiyuth Pramkeaw (Eds.)

Trends in Artificial Intelligence: PRICAI 2016 Workshops

PeHealth 2016, I3A 2016, AIED 2016
AI4T 2016, IWEC 2016, and RSAI 2016
Phuket, Thailand, August 22–23, 2016
Revised Selected Papers

 Springer

Editors
Masayuki Numao
ISIR
Osaka University
Ibaraki, Osaka
Japan

Thanaruk Theeramunkong
Thammasat University
Bangkadi Muang Pathumthani
Thailand

Thepchai Supnithi
National Electronics and Computer
  Technology Center
Language and Semantic Technology Lab
Pathum Thani
Thailand

Mahasak Ketcham
Information Technology
King Mongkut's University of Technology
  North Bangkok
Bangkok
Thailand

Narit Hnoohom
Mahidol University
Nakhon Pathom
Thailand

Patiyuth Pramkeaw
Media Technology
King Mongkut's University of Technology
  Thonburi
Thonburi
Thailand

ISSN 0302-9743   ISSN 1611-3349 (electronic)
Lecture Notes in Artificial Intelligence
ISBN 978-3-319-60674-3   ISBN 978-3-319-60675-0   (eBook)
DOI 10.1007/978-3-319-60675-0

Library of Congress Control Number: 2017943854

LNCS Sublibrary: SL7 – Artificial Intelligence

© Springer International Publishing AG 2017
This work is subject to copyright. All rights are reserved by the Publisher, whether the whole or part of the material is concerned, specifically the rights of translation, reprinting, reuse of illustrations, recitation, broadcasting, reproduction on microfilms or in any other physical way, and transmission or information storage and retrieval, electronic adaptation, computer software, or by similar or dissimilar methodology now known or hereafter developed.
The use of general descriptive names, registered names, trademarks, service marks, etc. in this publication does not imply, even in the absence of a specific statement, that such names are exempt from the relevant protective laws and regulations and therefore free for general use.
The publisher, the authors and the editors are safe to assume that the advice and information in this book are believed to be true and accurate at the date of publication. Neither the publisher nor the authors or the editors give a warranty, express or implied, with respect to the material contained herein or for any errors or omissions that may have been made. The publisher remains neutral with regard to jurisdictional claims in published maps and institutional affiliations.

Printed on acid-free paper

This Springer imprint is published by Springer Nature
The registered company is Springer International Publishing AG
The registered company address is: Gewerbestrasse 11, 6330 Cham, Switzerland

Preface

Established in 1990, the Pacific Rim International Conference on Artificial Intelligence (PRICAI) is a series of biennial international events which concentrate research on AI theories, technologies, and their applications among Pacific Rim countries. As a major conference in the AI field, scholars and researchers biannually participate to exchange their ideas. This year in the PRICAI 2016 conference, a number of workshops were arranged to bring together researchers of various backgrounds to present, discuss, and explore the state of applying information technology in various aspects of learning. The volume contains the material published for the workshops held with the 14th Pacific Rim International Conference on Artificial Intelligence (August 22–26, 2016, in Phuket, Thailand, http://aiat.in.th/pricai2016/).

We accepted six workshop proposals with the goal of exploring specific issues across various themes. Each paper in these proceedings was peer-reviewed by international reviewers to ensure the highest quality work. Among 46 submitted papers, 36 are accepted to be presented at the workshops, of which 23 were regular papers and 13 short papers. After an intense discussion during the workshops, we selected 16 high-quality papers (34% of the 46 submitted papers) for this volume.

Finally, we would like to thank the Pacific Rim International Conference on Artificial Intelligence (PRICAI 2016) Executive Committees and program co-chairs for entrusting us with the important task of chairing the workshop program, thus giving us an opportunity to grow through this valuable academic learning experience. We also would like to thanks all workshop co-chairs for their tremendous and excellence work.

February 2017

Masayuki Numao
Thanaruk Theeramunkong
Thepchai Supnithi
Mahasak Ketcham
Narit Hnoohom
Patiyuth Pramkeaw

Organization

Organizing Committee

Honorary Co-chairs

Wai Kiang (Albert) Yeap	AUT University, New Zealand
Abdul Sattar	Griffith University, Australia
Hiroshi Motoda	Osaka University, Japan
Vilas Wuwongse	Mahidol University, Thailand
Somnuk Tangtermsirikul	Thammasat University, Thailand
Sarun Sumriddetchkajorn	NECTEC, Thailand
Pun Thongchunum	Prince of Songkla University, Thailand

General Co-chairs

Dickson Lukose	MIMOS Berhad, Malaysia
Thanaruk Theeramunkong	Thammasat University, Thailand

Workshop Organizers

PeHealth

Chuleerat Jaruskulchai	Kasetsart University, Thailand
Ornuma Thesprasith	Kasetsart University, Thailand
Rey-Long Liu	Tzu Chi University, Taiwan

I3A

Narit Hnoohom	Mahidol University, Thailand
Tanasanee Phienthrakul	Mahidol University, Thailand
Mingmanas Sivaraksa	Mahidol University, Thailand
Anuchit Jitpattanakul	KMUTNB, Thailand
Sakorn Mekruksavanich	University of Phayao, Thailand
Ghita Berrada	Twente University, The Netherlands
Remi Barillec	Aston University, UK
Rozlina Mohamed	University Technology Malaysia, Malaysia

AIED

Thepchai Supnithi	NECTEC, Thailand
Rachada Kongkrachandra	Thammasat University, Thailand
Tsukasa Hirashima	Hiroshima University, Japan

AI4T

Manabu Okumura	Tokyo Institute of Technology, Japan
Hidetsugu Nanba	Hiroshima City University, Japan
Kazutaka Shimada	Kyushu Institute of Technology, Japan
Fumito Masui	Kitami Institute of Technology, Japan
ThanarukTheeramunkong	Thammasat University, Thailand

IWEC

Merlin Teodosia Suarez	De La Salle University, Philippines
The Duy Bui	Vietnam National University, Vietnam
Ma Mercedes Rodrigo	Ateneo de Manila University, Philippines
Masayuki Numao	Osaka University, Japan

RSAI

Vincent Shin-Mu Tseng	National Cheng Kung University, Taiwan
Mahasak Ketcham	KMUTNB, Thailand
Thaweesak Yingthawornsuk	KMUTT, Thailand
Narit Hnoohom	Mahidol University, Thailand
Patiyuth Pramkeaw	KMUTT, Thailand
Pokpong Songmuang	Thammasat University, Thailand

Program Committee

Thanaruk Theeramunkong	Thammasat University, Thailand
Masayuki Numao	Osaka University, Japan
Boonserm Kijsirikul	Chulalongkorn University, Thailand
Thepchai Supnithi	NECTEC, Thailand
Akihiro Kashihara	University of Electro-Communications, Japan
Anantaporn Srisawat	KMITL, Thailand
Anuchit Jitpattanakul	KMUTNB, Thailand
Apichat Suratanee	KMUTNB, Thailand
Chalermsub Sangkavichitr	KMUTT, Thailand
Chi-Jen Lin	National Taiwan University, Taiwan
Dennis Reidsma	University of Twente, The Netherlands
Dittaya Wanvarie	Chulalongkorn University, Thailand
Ekawat Chaowicharat	Thammasat University, Thailand
Elisabetta Bevacqua	Lab-STICC, CERV-ENIB, France
Eriko Aiba	University of Electro-Communications, Japan
Fumito Masui	Kitami Institute of Technology
Hidetsugu Nanba	Hiroshima City University
Isao Ono	Tokyo Institute of Technology, Japan
Iwan Dekok	University of Twente, The Netherlands
Jerome Urbain	University of Mons, Belgium
Jiradej Ponsawat	KhonKaen University, Thailand

Joseph Beck	Worcester Polytechnic Institute, USA
Kazutaka Shimada	Kyushu Institute of Technology
Khiet Truong	University of Twente, The Netherlands
Kitiporn Plaimas	Chulalongkorn University, Thailand
Komate Amphawan	Burapha University, Thailand
Kritya Bunjongjit	Mahidol University, Thailand
Magalie Ochs	Telecom Paris Tech, France
Mahasak Ketcham	KMUTNB, Thailand
Manabu Okumura	Tokyo Institute of Technology
Masashi Inoue	Yamagata University, Japan
Matinee Kiewkanya	Chiang Mai University, Thailand
Mingmanas Sivaraksa	Mahidol University, Thailand
Narit Hnoohom	Mahidol University, Thailand
Narumol Chumuang	KMUTNB, Thailand
Niwat Srisawasdi	Khon Kaen University, Thailand
Noriko Otani	Tokyo City University, Japan
Ornprapa Pummakarnchana Robert	Silpakorn University, Thailand
Patcharin Panjaburee	Mahidol University, Thailand
Patiyuth Pramkeaw	KMUTT, Thailand
Phaisarn Jeefoo	University of Phayao, Thailand
Pisit Phokharatkul	Mahidol University, Thailand
RadoslawNiewiadomski	Telecom Paris Tech, France
Rajeswari Matam	Birmingham Children's Hospital, UK
Randa Herzallah	Aston University, UK
Rangsipan Marukatat	Mahidol University, Thailand
Ryan Baker	Columbia University, USA
Sakkayaphop Pravesjit	University of Phayao, Thailand
Sakorn Mekruksavanich	University of Phayao, Thailand
Sanparith Marukatat	NECTEC, Thailand
Seta Kazuhisa	Osaka Prefecture University, Japan
Sidney D'Mello	University of Notre Dame, USA
Sittichai Choosumrong	Naresuan University, Thailand
Sunisa Rimcharoen	Burapha University, Thailand
Tanapon Jensuttiwetchakul	KMUTNB, Thailand
Tanasanee Phienthrakul	Mahidol University, Thailand
Thatsanawan Soonklang	Silpakorn University, Thailand
Thaweesak Yingthawornsuk	KMUTT, Thailand
Thittaporn Ganokratanaa	Chulalongkorn University, Thailand
Thomas Bermudez	Ernst & Young, UK
Tomoko Kojiri	Kansai University, Japan
Tsukasa Hirashima	Hiroshima University, Japan
Ulrich Hoppe	University of Duisburg-Essen, Germany
Wiwit Suksangaram	PBRU, Thailand
Worawut YimYam	PBRU, Thailand

Contents

PeHealth 2016: Workshop on eHealth Mining

Exploring the Distributional Semantic Relation for ADR and Therapeutic Indication Identification in EMR

Siriwon Taewijit[1,2(✉)] and Thanaruk Theeramunkong[1]

[1] Sirindhorn International Institute of Technology, Thammasat University,
Klong Luang, Pathum Thani, Thailand
thanaruk@siit.tu.ac.th
[2] Japan Advanced Institute of Science and Technology, Nomi, Ishikawa, Japan
siriwont@jaist.ac.jp

Abstract. Extraction of relations and their semantic relations from a clinical text is significant to comprehend the actionable harmful and beneficial events between two clinical entities. Particularly to implement drug safety surveillance, two simplest but most important semantic relations are *adverse drug reaction* and *therapeutic indication*. In this paper, a method to identify such semantic relations is proposed. A large scale of nearly 1.6 million sentences over 50,998 discharge summary from Electronic Medical Records were preliminary explored. Our approach provided the three main contributions; (i) Electronic Medical Records characteristic exploration; (ii) OpenIE examination for clinical text mining; (iii) automatic semantic relation identification. In this paper, the two complementary information from public knowledge base were introduced as a comparative advantage over expert annotation. Then the set of relation patterns were qualified with *0.05* significant level. The experimental results show that our method can identify the common *adverse drug reaction* and *therapeutic indication* with the high lift value. Additionally, a novel *adverse drug reaction* and alternative drug for a specific symptom therapy are reported to support the comprehensive further drug safety surveillance. The paper clearly illustrates that our method is not only effortless from expert annotation, automatic pattern-specific semantic relation extraction, but also effective for semantic relation identification.

Keywords: Adverse drug reaction · Electronic Medical Records · Semantic relation extraction · Therapeutic indication · Text mining

1 Introduction

Exponential growth of Electronic Medical Records (EMR), with replacement of paper-based records enables us to utilize information in effective and efficient ways. Recently several works have been focused on semantic relation extractionfrom EMR to conspicuously benefit to *drug-symptom,* network [1, 2] for comprehensive drug safety surveillance [3–5]. Basically, there are two simply complementary semantic (meaning) relations are existing along with the network; *drug-induce-symptom* for the harmful perspective and *drug-treat-symptom* for the beneficial one. The former is namely as

© Springer International Publishing AG 2017
M. Numao et al. (Eds.): PRICAI 2016 Workshops, LNAI 10004, pp. 3–15, 2017.
DOI: 10.1007/978-3-319-60675-0_1

adverse drug reaction (ADR) and the latter refers to therapeutic indication. To deal with EMR, text preprocessing (e.g. sentence boundary detection (SBD), name entity recognition (NER), etc.) and the machine-readable form for text representation are the mandatory tasks which highlight the challenges.

On the one hand, in order to explore associations between arbitrary two entities in huge EMR contexts, the relation extraction is a fundamental process. There are two main paradigms for relation extraction [6]. Firstly, the traditional relation extraction primarily recognises a relation-specific between two or more entities in text. The process is fallen into drawback of labouring tasks due to hand-labeled training and infeasible less efforts when shifting to a new relation. The cost and time-consuming of this manual tasks are linearly expensive along with the number of relations. Another paradigm, open information extraction (OpenIE) [7], is rather new and feasible for a large scale corpora due to relation independent extraction. Hence, a vast number of diversity of relational tuples between arbitrary two clinical entities are extracted simultaneously. Putting semantic into these relational tuples can further enhance the comprehensive regarding the actionable harmful and beneficial events. Practically, the semantic relations [8–10] are not always expressed with explicit words such as *treat* or *cause* etc. Mostly, they are also frequently expressed with combined and complex expressions [11] and need interpretation. In our research contexts, the semantic relation identification in narrative text from EMR corresponds to automatically annotate the relation type (e.g. *ADR, indication*). Considering on the one of our discovered relation, "*be hold in*" and "*be appropriately controlled with*", they can be annotated into *ADR* and *indication* semantic relations respectively.

In this work, we provided three contributions; (i) to explore the characteristic of discharge summary from EMR; (ii) to examine the powerful of OpenIE on narrative text; (iii) to automatically identify semantic relation. We initially analyzed a huge number of nearly 1.6 million sentences over 50,998 discharge summary from EMR. The two rich contexts from the brief hospital course (BHC) and the history of present illness (HPI) sections in discharge summary were selected for investigation. Then the rather simple but efficient method by exploring distributional semantic relation was employed in order to capture the key pattern-specific semantic relation for *ADR* and *indication* identification. To avoid suffering from hand-labeled training, the comparatively existing knowledge base from SIDER and DrugBank were enriched as temporary label. Our hypothesis is that the pattern-specific semantic relation is significantly related to its label. The conditional entropy was computed to examine the uncertainty of semantic relation given a pattern. Lately, the key pattern-specific semantic relation was qualified by hypothesis testing of contingency tables with *0.05* significant level. We organise the remaining of this paper into five sections: the related work is given in the next section. The text mining for data preprocessing and distributional semantic relation exploration are described in Sect. 3. The experimental results and conclusion are summarised into Sects. 4 and 5 respectively.

2 Related Work

Detection of underlying relations to further comprehend *drug-symptom* network, multidisciplinary approaches are extensive study. Traditional approach, co-occurrence statistic is favour for decade [12–14] due to simplest and less effort. The method relies on the co-occur of a pair of drug and symptom entities in the specified boundary such as with in a window size, sentence, abstract, paragraph, or document. Unfortunately, this method is loosely to capture the true semantic relation such as *drug-induce-symptom* or *drug-treat-symptom*.

The considering of distinct between two complementary semantic relations of *ADR* and *indication* has been the significant of comprehensive *drug-symptom* network. Wang et al. [15] incorporated omic data (i.e. chemical structures and protein targets) and the two complementary semantic relations. The *ADR* and *indication* semantic relations are interchangeable as a feature representation for themselves predictive model, for example, *ADR* with omic data as feature representation to predict *indication*, and *vice versa*. Then two interdependent models were constructed to estimate the probability of *ADR* and *indication* by logistic regression. Recently, the preliminary study of ADR and therapeutic indication on social media, Segura-Bedmar et al. [16] employed co-occurrence of *drug-symptom* pairs by varying n window size from 10 to 50 on 400 Spanish user comments. Hence, the semantic relation was assigned based on the appearance of *drug-symptom* pairs according to its sections describing (ADR or indication) which was derived from drug package leaflets.

Lately, two works but complementary are reported by Xu et al. The former [17] aimed to derive new drug therapeutic indication for drug repurposing by explore *drug-symptom* pairs from 20 million MEDLINE abstracts. To learn the drug indication pattern, *drug-symptom* pairs were extracted from Clinicaltrials.gov, then, the contexts between all extracted pairs were examined for *indication* pattern. Latter work [18] employed dependency parse tree to retrieve ADR-specific syntactic patterns for ADR detection. The large scale of 119 million MEDLINE sentences were investigated. Different from previous work, the SIDER knowledge base was deployed to derive *drug-symptom* pairs regarding ADR. Finally, all extracted patterns were ranked based on their associated pattern scores and co-occurrence frequencies. Then, the manual selection process was manipulated in order to remove irrelevance patterns and used to retrieve unknown *drug-symptom* pairs. The example patterns regarded to *ADR* and *indication* semantic relations are *"induced"*, *"associated"*, *"related"*, *etc.* and *"in"*, *"for the treatment of"*, *"in the management of"*, *etc.* respectively.

3 Materials and Methods

The entire experimental process was divided into four main tasks: (i) data pre-processing; (ii) information extraction; (iii) the key pattern-specific semantic relation identification; (iv) semantic relation inference. Our proposed method was shown in Fig. 1.

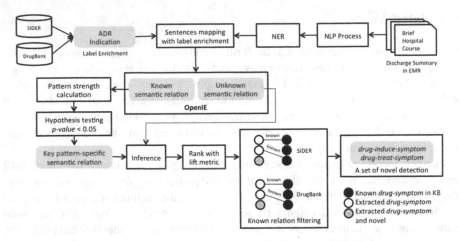

Fig. 1. The framework for semantic relations identification. The two public knowledge bases are comparative enrichment as labeled relations rather than domain expert annotation. The distributional semantic relations are examined in order to capture pattern–specific semantic relation, then the key pattern-specific semantic relation was qualified by hypothesis testing of contingency tables with *0.05* significant level. Lastly, the inference process was employed to derive novel semantic *drug-symptom* pairs.

3.1   Data Preprocessing

In order to explore the relation between arbitrary two clinical entities to comprehend harmful and beneficial actions corresponding *drug-symptom* network, we set up the experimental study by utilizing the public source of EMR from MIMIC-III [19]. The database is introduced by *National Institute of Biomedical Imaging and Bioengineering* and available at *PhysioNet*[1]. Two sections of BHC and HPI over 58,000 observed hospital admissions were extracted. The preprocess tasks were a considerable prerequisite for further text analysis.

The sentence boundary detection (SBD) is comparatively fundamental task in Natural Language Processing (NLP), but significantly important regarding the text quality. There were a number of noise-prone in the narrative text and essentially needed to manipulate. For instance, in MIMIC-III, the period (".") in the narrative text can be employed as sentence boundary marker, abbreviations, level number in laboratory results, medication dosage, etc. The discontinuous text over a line is an irritated ill-form as well. The capital letter along with the punctuation mark such as colon (":") for content section expression also increased the challenges. Figure 2 depicted a natural format of the narrative text from MIMIC-III. We tackled such noise-prone by developing an in-house SBD rather than utilizing *a state-of-the-art* method to compatible with the narrative text from EMR. The heuristic patterns were predefined to carry out the unregulated linguistic form. Eventually, nearly 1.6 million sentences were extracted from BHC and HPI sections in the discharge summary.

---

[1] https://mimic.physionet.org.

In addition, highly accurate semantic relations identification is strongly related to clinical entities extraction. This is a common task in text mining, which corresponds to NER in Information Retrieval. We accomplished NER for a given narrative text using notable MetaMap tool [20]. The tool recognizes a clinical term in the given narrative text and results its standard term provided by Unified Medical Language System (UMLS). Our post processing was employed on *the out-of-the-box* MetaMap results to overcome the ambiguous NER. The two semantic group codes of *cherriical* and *disorder* were considered for drug and symptom entities respectively. The summary of statistical number was placed in Table 1.

Table 1. The statistical number of contexts derived from the brief hospital course (BHC) and the history of present illness (HPI) in EMR.

	EMR Corpus			Knowledge base (KB)		
	Total	BHC	HPI	Total	BHC	HPI
Discharge summary	49,271	36,907	49,092	–	–	–
Sentences	1,580,628	980,795	599,833	–	–	–
Sentences *(drug-symptom)*	218,135	124,074	94,061	–	–	–
Sentences/document						
+min/max	1/251	1/248	1/11	-	-	-
+avg./std.	31/23.7	26/19.1	8 12/8.7	-	-	-
Drug terms *(matched to KB)*	3,231	2,637	2,184	–	–	–
Symptom terms *(matched toKB)*	9,960	7,648	7,406	–	–	–
Relation extraction						
All open relations	1,210,501	675,664	534,837	–	–	–
Drug terms	1,142	639	977	192	168	78
Symptom terms	3,080	2,143	2,111	190	171	78
drug-induce-symptom	77,652	43,088	34,564	589	480	109
drug-treat-symptom				732	553	179

ARF: On arrival the patient's creatinine was 1.5 (up from baseline 1.2-1.3). Over the course of his stay his creatinine increased as high as 2.2. This was believed to be multifactorial, likely due to decreased renal perfusion in setting of his arrest, infection, and possibly due to gentamycin, although his trough level never exceeded 1.9. He was followed by the renal team during his stay and had a bland urine, negative for eosinophils. His creatinine remained stable at about 2.2 for 5 days before discharge. His outpatient ramipril was held in the setting of acute renal insufficiency and may be restarted as an outpatient when his renal function is more stable.

Fig. 2. An example of narrative text in discharge summary from MIMIC-III.

3.2   Information Extraction

A new paradigm OpenIE is a generalization of typical information extraction (IE). OpenIE provides the potential effort to deal with the large-scale corpora without manual tagging of relations [21], while the traditional one fully requires precisely target relation beforehand. Early of OpenIE [22] aimed to extract an unknown relation in advance on highly scalable Web corpus. The evident achievements on web mining let to an extensive paradigm shift in medical text mining. Recently, the Stanford CoreNLP developed OpenIE [23] in order to reduce a large pattern set for canonical sentences and excerpt self-contained clauses from longer sentences as well.

In our work, given a set of sentences *s* from the preprocessing process, the Stanford OpenIE was carried out to examine the powerful on clinical text mining (Fig. 3). As the results, 1.2 million of the massive number of domain independent relational tuples < *arg1, pattern, arg2*> was reported. Unfortunately, not all but some of tuples can represent *drug-symptom* relationship. We, hence, filtered irrelevance relations and derived 77,652 *drug-symptom* relational tuples.

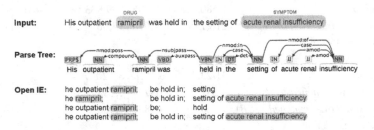

Fig. 3. An illustration of toy example of relation extraction using the Stanford OpenIE. The top of the figure is a preprocessed input. The sentence contains *ramipril* drug (concept id C0072973) and *acute renal insufficiency* symptom (concept id C0022660). The grammatical structure from parse tree of the given sentence was placed in the middle of the figure. The set of clauses from the Stanford OpenIE was shown in the bottom of the figure. It contains diverse self-contained clauses, which preserves both of syntactic and semantic entailing by the original sentence.

3.3   The Key Pattern-Specific Semantic Relation Extraction

The identification of semantic relation between arbitrary two clinical entities of drug and symptom is necessary for further harmful and beneficial analysis. Our hypothesis is that a mediated relation between drug and symptom entities should specific to its semantic relation (e.g. *ADR, indication*). This relation dependency analysis needed the known pair of *drug-induce-symptom* and *drug-treat-symptom* beforehand. To avoid the expensive of expert domain acquisition, the public and freely accessible knowledge bases from DrugBank and SIDER were comparatively incorporated as such sources of annotated data. While SIDER was utilized as harmful relation *(ADR)* and DrugBank was obtained for beneficial relation *(indication)* subsequently. The considering of annotated data from knowledge base are well-known as population base while our corpus is instance level.

The utilization of the fact from knowledge base, *firstly*, the known semantic relations were temporarily annotated into each relational tuple derived from OpenIE. For example, *drug-symptom* pair of *amiodarone-hypotension* existed in SIDER. We, then, annotated the label of *ADR* into all relational tuples that contained *amiodarone* drug and *hypotension* symptom. Conversely, *drug-symptom* pair of *lahetalol-hypertension* existed in DrugBank, therefore, the label of *indication* was annotated into all relational tuples that contained *labetalol* drug and *hypertension* as well. However, when we considered on the mediated context between *drug-symptom* pair for each sentence, the temporary annotation might not be always true.

Secondly, the pattern-specific semantic relation was investigated. Based on our hypothesis, we observed the distribution of the extracted pattern along with its semantic relation. The conditional entropy (Eq. 1) was examined to quantify the degree of uncertainty for each pattern. After that, the pattern strength was obtained by conditional entropy adjustment (Eq. 2), the higher score, the stronger pattern strength.

Finally, we filtered out unreliable patterns by examining the hypothesis testing of association between the semantic relation given a specific extracted pattern. The statistical Fisher's exact test at *0.05* significant level was considered. The qualified patterns were resulted as the key pattern-specific semantic relation. In summary, we derived 353 key pattern-specific semantic relation; 216 for *ADR;* 137 for *indication*. Difference from the previous study by Xu et al. [17, 18], our proposed method is automatic identification, non redundant, and feasible for large amount of the key pattern-specific semantic relation derivation. We exhibited the ranking of the key pattern-specific semantic relation in Fig. 4. Additionally, Table 2 illustrated the top 5 key pattern-specific semantic relation and sample sentences for *drug-induce-symptom* and *drug-treat-symptom* relations.

Given an extracted pattern x_i and semantic relation $y_i \in Y$ where as $Y = \{ADR,$ *indication}*, the conditional entropy and pattern strength were defined as follows:

$$H(Y|X = x_i) = -\sum_{j=1}^{C} P(y_j|x_i) \log_2 P(y_j|x_i) \qquad (1)$$

$$P_{strength}(X = x_i) = (1 - H(Y|X = x_i))(P(y_j|x_i) - (1 - P(y_j|x_i))) \qquad (2)$$

3.4 Semantic Relation Inference

The extracted relational tuples derived from OpenIE were queried through all key pattern-specific semantic relation in order to infer the semantic relation. Then we computed the lift metric to evaluate the likelihood of a *drug-symptom* pair against the co-occurrence by chance. The lift value over than 1 implies the stronger association between drug and symptom over the chance (lift = 1)

$$lift(drug, symptom) = \frac{P(drug, symptom)}{P(drug)P(symptom)} \qquad (3)$$

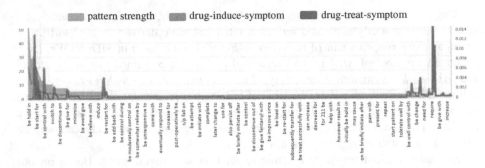

Fig. 4. The set of key pattern-specific semantic relation is ranked by the pattern strength (grey area), the higher score, the stronger pattern strength. The frequency of *drug-induce-symptom* and *drug-treat-symptom* are exhibited with red and green lines accordingly. (Color figure online)

4 Experimental Results

We investigated the performance of our proposed method and reported into two parts; (i) analysis of the characteristics of discharge summary; (ii) evaluation of the key pattern-specific semantic relation.

4.1 Analysis of the Characteristics of Discharge Summary

In order to comprehend the data characteristics of EMR, nearly 1.6 sentences of 50,998 discharge summary was explored. The seven main sections were narrated in discharge summary (Fig. 5). The maximum number of sentences was located in the BHC section. On the one hand, the HPI and the discharge medications sections contained equally number of sentences, however, contexts in the discharge medications were mostly written as a list of drug prescription regardless symptom description. Unfortunately, not all sections described the purpose of drug prescription or adverse reaction from drug usage, which can contribute to further research of *drug-symptom* network. Our work initially investigated on BHC and HPI sections that their information were closely related to *drug-induce-symptom* and *drug-treat-symptom*.

From the Table 1, the BHC section contained the number of sentences more than the HPI around 1.3 times and the average sentences per document are 26 and 12 respectively. Then drug and symptom entities were extracted, it was astonished that the both sections contained equally the same number of drug and symptom entities. This is because the HPI section is permeated with clinical contents that are directly related to patient's symptom and remedy, while, the BHC narrates patient health status before, during, and after admission including treatment courses in more details. Afterwards, the relational tuples from all sentences were extracted by using the Stanford OpenIE. We found that nearly 6.4% (77,652) of all extracted relational tuples contained both drug and symptom entities in the same sentence. In contrast, the remaining (93.6%) of extracted relational tuples contained only drug, only symptom, or not related to *drug-symptom relation*.

Table 2. Top 5 of the extracted key pattern-specific semantic relation and the example sentences from EMR.

Top 5	Key patterns	Example sentences
Drug-induce-symptom(ADR)		
1	be hold in	His outpatient <u>ramipril</u> was held in the setting of <u>acute renal insufficiency</u>
2	contribute to	<u>Morphine</u> contributed to <u>urinary retention</u> as seen in high PVR so foley placed
3	be think	His <u>rash</u> was thought secondary to the <u>nafcillin</u>
4	improve with	Patient's <u>dysphagia</u> improved with iv <u>pantoprazole</u>
5	cause	<u>Propofol</u> caused mild <u>hypotension</u> to 95
Drug-treat-symptom (indication)		
1	continue	<u>Depression</u> - continue outpatient <u>fluoxetine</u>
2	be start for	<u>Phenylephrine</u> drip was started for <u>hypotension</u>
3	be on	<u>RHEUMATOID ARTHRITIS</u>: The patient is on <u>Methotrexate</u> at home
4	be control with	The patient's <u>blood pressure</u> was controlled with <u>labetalol</u>
5	be add for	<u>Norepinephrine</u> was later added for persistent <u>hypotension</u>

Finally, nearly 1.7%(1,321) from 77,652 drug-symptom relation were derived as temporary semantic relations corresponding known relations from SIDER and Drug-Bank, hence, they were qualified for the key pattern-specific semantic relation extraction.

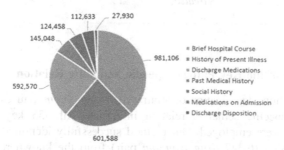

Fig. 5. The number of sentences of the seven sections in discharge summary from Mimic-III.

Moreover, our key pattern-specific semantic relation were compared with the one from the two studies of Xu et al. The key pattern phrases from our method are rather different from Xu et al. due to different writing style (Table 3). The study of Xu et al. via MEDLINE corpus provided the shorter phrases such as only *verb* or *preposition*, and not express auxiliary verb. Our key pattern phrases derived from EMR using the Stanford OpenIE are longer with more complete phrase such as *"be in"* rather than *"in"* or *"well control with"* rather than *"with"* etc. Mostly, the extracted key pattern phrases are fully contained *verb* and followed by *preposition*, which benefits to relation comprehension and expose theirs semantic corresponding *ADR* and *indication*. Moreover, unlike the report from two studies of Xu et al., our proposed method certainly discriminated patterns between *ADR* and *indication*, so no overlapped patterns were found. The uncertain pattern regarding the overlapping is well-known as the cause of false positive.

Table 3. The key pattern-specific semantic relation comparison between our proposed method, which are derived by Stanford OpenIE, and the two studies of Xu et al.

	ADR		Indication	
	Drug-symptom	Symptom-drug	Drug-symptom	Symptom-drug
Our method	*be hold in*	*be in*	*continue*	*be continue on*
	be hold	*improve with*	*be start for*	*be control with*
	contribute to	*be think*	*be add for*	*be on*
	be hold for	*be attribute to*	*be initiate for*	*well control*
	cause	*think*	*be give for*	*with*
	be discontinue	*feel*	*be continue for*	*continue on*
	be hold give	*hold for*	*be restart for*	*be control on*
	...	...	...	*be treat with*
				...
Xu et al. [17, 18]	*_induced*	*induced_by*	*in*	*with*
	induced	*after*	*for the treatment of*	*were treated*
	_associated	*caused by*	*treatment of*	*with*
	_related	*following*	*in the management of*	*to*
	induces	*produced by*	*_resistant*	*after*
	caused	*after treatment*	*in a patient with*	*during*
	developed	*in patients*	*to treat*	*in*
	...	*treated*	...	*associate with*
		...		...

4.2   Evaluation of the Key Pattern Specific Semantic Relation

The semantic relation identification of arbitrary *drug-symptom*, pair can be derived by the key pattern-specific semantic relation inference. All 353 key pattern-specific semantic relation were employed. Our method successfully identified approximate to five times increasing (6,347 *drug-symptom* pair) from the known relation. We randomly selected four sets of *Metoprolol-symptom* and *drug-Hypotension* for *ADR* semantic relation, and *Amiodarone-symptom* and *drug-Pneumonia* for *indication* semantic relation. The lift metric was used to evaluate the likelihood of a *drug-symptom* pair against co-occurrence by chance (Table 4).

From FDA prescribing information, *Metoprolol* drug is indicated to treat chest pain, hypertension, and prevents heart attack. The key pattern-specific *ADR* semantic relation was used to query *Metoprolol-symptom* pair. We found that the frequent and common *ADR* caused by *Metoprolol* such as *AV block, Heart block, Hypotension* has the higher lift value, while the no frequent information or rare such as *rash* and *tachycardia* provided the small number over chance. The novel *Metoprolol-Pulmonary* pair which is not existing in the KB, was derived from our method. The RxList[2], the premier Internet Drug Index resource, was used to verify the identified semantic relation. We found that *dyspnea of pulmonary origin* was reported as *ADR* of *Metoprolol* drug. Another one, the novel *Metoprolol-Kidney failure* pair was reported by Mayo Clinic[3] as the possible *ADR* for long-term treatment. Identically, *drug-Hypotension* pairs regarding the key pattern-specific *ADR* semantic relation were queried. According to the fact from SIDER, hypotension is a common ADR of *Losartan* and *Metoprolol* drugs, thus, the lift value is placed in the high order. The *Ciprofloxacin-Hypotension* has the lower lift value relevance to the rare and uncommon ADR.

In the contrast, the lift value of *therapeutic indication* semantic relation is totally different from *ADR*. In which, the *indication* semantic relation provides the bigger number of lift due to the drug prescribing regularly with known indication for hospitalisation. Considering on *Amiodarone-symptom* pair, all relevance symptoms have the high lift score because *Amiodarone* drug is indicated to treat heart rhythm disorders. Another one, *Pneumonia* symptom, this symptom is an infection of the lungs and it can be caused by bacteria, viruses or fungi. All of the drugs relevance *Amiodarone-symptom* pair, that are listed in the Table 4, are indicated to treat bacterial infections. However, here in, only *Azithromycin* and *Cefepime* which were reported in DrugBank database. Our method can derive the alternative drug therapy (*Levofloxacin* and *ceftriaxone*) for *Pneumonia* with the lift of 24.95 and 14.40 respectively.

Table 4. The semantic identification regarding the key pattern-specific semantic relation. The lift value is used to evaluate the association between *drug-symptom* pair through our semantic relation identification over co-occurrence by chance. The *KB* column is marked yes, if *drug-symptom* pair exists in our knowledge base that are extracted from SIDER or DrugBank, and *vice versa*.

Metoprolol-induce-symptom	Lift	KB	Drug-induce-Hypotension	Lift	KB
AV block	11.91	yes	Los art an	5.67	yes
Heart block	11.91	yes	Metoprolol	5.24	yes
Pulmonary disease	5.96	new	Propofol	4.99	yes
Hypotension	5.24	yes	Garvedilol	4.86	yes
Asystolic	3.40	yes	Imdur	4.86	new
Sepsis	2.70	new	Furosemide	3.60	yes
Kidney failure	2.38	new	Nadolol	3.24	yes
Tachycardia	2.09	yes	Atenolol	3.13	yes
Rash	1.54	yes	Ciprofloxacin	2.05	yes
Amiodarone-treat-symptom	Lift	KB	Drug-treat-Pneumonia	Lift	KB
Ventricular arrythmia	28.40	yes	Levofloxacin	24.95	new
Rhythm	19.88	yes	Azithromycin	17.43	yes
Ventricular tachycardia	17.04	yes	Ceftriaxone	14.40	new
Atrial fibrillation	11.09	yes	Cefepime	11.20	yes

---

[2] http://www.rxlist.com/.

[3] http://www.mayoclinic.org/.

5 Conclusions

We introduced the framework to identify semantic relation of *ADR* and *indication* from a large scale of narrative text from EMR. From our initial investigation, nearly 1.6 million sentences were examined by enrichment labeled data from the two sources of knowledge base SIDER and DrugBank. Consequently, the notable Stanford OpenIE was carried out to retrieve mediated relations between two entities of *drug* and *symptom* as a result of tuple relation *<arg1, pattern, arg2>*. Henceforth, the conditional entropy with *0.05* significant level was presented to capture the pattern strength and automatically qualify the key pattern-specific semantic relation. Furthermore, the key pattern-specific semantic relation inference was employed in order to identify the semantic relation of new *drug-symptom* pair. Lastly, the lift metric was computed to measure likelihood of semantic association of *drug-symptom* pair. To derive the novel *drug-induce-symptom* pair and *drug-treat-symptom*, we filtered out the *drug-symptom* pair corresponding the known relations from SIDER and DrugBank respectively.

However, our method has some limitations that needs to be improved such as the low rate of recall due to small numbers of the key pattern-specific semantic relation, and the precision of drug and symptom NER. Moreover, OpenIE can retrieve diversity of relational tuples, but the method fails to discover partial or incomplete sentence especially the absent of *verb*, which is the natural pattern of narrative text in EMR.

From the experimental results, our work is not only effective and scalable for semantic relation identification, less expensive for expert annotation, but also promising framework to discover a novel harmful and beneficial drug therapeutic indication. This preliminary investigation of the utilization from EMR indicated that our contribution can support the further research of drug safety surveillance and drug repurposing as a screening method by systematic way.

Acknowledgment. Authors thank for the research environment, which was supported by National Research University project (NRU) and the Thammasat center of Excellence (CILS), Thailand.

References

1. Xu, X., Zhang, C., Li, P., Zhang, F., Gao, K., Chen, J., Shang, H.: Drug-symptom networking: Linking drug-likeness screening to drug discovery. Pharmacol. Res. **103**, 105–113 (2016)
2. Zhang, Y., Tao, C., Jiang, G., Nair, A.A., Su, J., Chute, C.G., Liu, H.: Network- based analysis reveals distinct association patterns in a semantic medline-based drug-disease-gene network. J. Biomed. Semant. **5**, 33 (2014)
3. Karlsson, I., Zhao, J., Asker, L., Boström, H.: Predicting adverse drug events by analyzing electronic patient records. In: Peek, N., Marín Morales, R., Peleg, M. (eds.) AIME 2013. LNCS, vol. 7885, pp. 125–129. Springer, Heidelberg (2013). doi:10.1007/978-3-642-38326-7_19

4. Park, M.Y., Yoon, D., Lee, K., Kang, S.Y., Park, I., Lee, S.-H., Kim, W., Kam, H.J., Lee, Y.-H., Kim, J.H., et al.: A novel algorithm for detection of adverse drug reaction signals using a hospital electronic medical record database. Pharmacoepidemiol. Drug Saf. **20**(6), 598–607 (2011)
5. Sohn, S., Kocher, J.-P.A., Chute, C.G., Savova, G.K.: Drug side effect ex-traction from clinical narratives of psychiatry and psychology patients. JAMIA **18**(Suppl.), 144–149 (2011)
6. Banko, M., Etzioni, O., Center, T.: The tradeoffs between open and traditional relation extraction. ACL **8**, 28–36 (2008)
7. Banko, M., Cafarella, M.J., Soderland, S., Broadhead, M., Etzioni, O.: Open information extraction for the web. IJCAI **7**, 2670–2676 (2007)
8. Lobner, S.: Understanding semantics. Routledge (2002, 2013)
9. Hurford, J.R., Heasley, B., Smith, M.B.: Semantics: A Coursebook. Cambridge University Press, New York (1983, 2007)
10. Lyons, J.: Linguistic Semantics: An Introduction. Cambridge University Press, Cambridge (1995)
11. Abacha, A.B., Zweigenbaum, P.: Automatic extraction of semantic relations between medical entities: a rule based approach. J. Biomed. Semant. **2**(5), 1 (2011)
12. Wang, X., Hripcsak, G., Markatou, M., Friedman, C.: Active computerized pharmacovigilance using natural language processing, statistics, and electronic health records: a feasibility study. J. Am. Med. Inform. Assoc. **16**(3), 328–337 (2009)
13. Wang, X., Hripcsak, G., Friedman, C.: Characterizing environmental and phe- notypic associations using information theory and electronic health records. BMC Bioinf. 10(Suppl. 9), S13 (2009)
14. Chen, E.S., Hripcsak, G., Xu, H., Markatou, M., Friedman, C.: Automated acquisition of disease-drug knowledge from biomedical and clinical documents: an initial study. J. Am. Med. Inform. Assoc. **15**(1), 87–98 (2008)
15. Wang, F., Zhang, P., Cao, N., Hu, J., Sorrentino, R.: Exploring the associations between drug side-effects and therapeutic indications. J. Biomed. Inform. **51**, 15–23 (2014)
16. Segura-Bedmar, I., De La Pena, S., Martinez, P.: Extracting drug indications and adverse drug reactions from Spanish health social media. In: Proceedings of the 2014 Workshop on Biomedical Natural Language Processing, pp. 98–106. ACL (2014)
17. Xu, R., Wang, Q.: Large-scale extraction of accurate drug-disease treatment pairs from biomedical literature for drug repurposing. BMC Bioinf. **14**, 181 (2013)
18. Xu, R., Wang, Q.: Automatic construction of a large-scale and accurate drug-side-effect association knowledge base from biomedical literature. J. Biomed. Info. **51**, 191–199 (2014)
19. Goldberger, A.L., Amaral, L.A.N., Glass, L., Hausdorff, J.M., Ivanov, P.C., Mark, R.G., Mietus, J.E., Moody, G.B., Peng, C.K., Stanley, H.E.: PhysioBank, PhysioToolkit, and PhysioNet: Components of a new research resource for complex physiologic signals. Circulation **101**(23), e215–e220 (2000)
20. Aronson, A.R.: Effective mapping of biomedical text to the UMLS Metathesaurus: The MetaMap program. In: Proceedings of the AMIA Annual Symposium, pp. 17–21 (2001)
21. Soderland, S., Roof, B., Qin, B., Xu, S., Etzioni, O.: Adapting open information extraction to domain-specific relations. AI Mag. **31**(3), 93–102 (2010)
22. Etzioni, O., Banko, M., Soderland, S., Weld, D.S.: Open information extraction from the web. Commun ACM. **51**(12), 68–74 (2008)
23. Angeli, G., Premkumar, M.J., Manning, C.D.: Leveraging linguistic structure For open domain information extraction. In: Proceedings of the Association of Computational Linguistics (ACL), pp. 26–31 (2015)

I3A 2016: Workshop on Image, Information and Intelligent Applications

Desktop Tower Defense Is NP-Hard

Vasin Suttichaya[✉]

Department of Computer Engineering, Faculty of Engineering,
Mahidol University, Salaya, Thailand
vasin.sut@mahidol.ac.th

Abstract. This paper proves the hardness of the Desktop Tower Defense game. Specifically, the problem of determining where to locate k turrets in the grid of size $n \times n$ in order to maximize the minimum distance from the starting point to the terminating point is shown to be NP-hard. The proof applied to the generalized version of the Desktop Tower Defense.

Keywords: Graph theory · Hamiltonian path · Complexity

1 Introduction

Tower Defense (TD) is a type strategy video game that concentrates on protecting some part of the territory from waves of enemies, also known as *"creeps"*. Enemies always appear at the entrance and attempt to walk to the exit point. The player must plan defensive strategies for protecting their bases, usually achieved by placing turrets alongside the enemy's road. Winning TD can be a painful task since the player needs to concurrently optimize many factors, such as resource, location, and enemy's abilities.

Desktop Tower Defense (DTD) is a popular tower defense game. The major difference between classic TD and DTD is the player's ability to control the enemy's path from the starting point to the exit point. DTD allows the player to build turrets on any positions on the map. Turrets can be used as walls for blocking enemies, and force them to find the new shortest path to the exit point. The only limitation is the player cannot place turrets in the way such that they completely block the exit. Therefore, the best strategy for winning DTD is not only optimizing own resources, but also instantly extending the distance. The DTD's gameplay is illustrated in Fig. 1.

TD introduces many challenges in problem solving areas, such as resource allocation, and geometry problems. Therefore, it is interested by researchers in the artificial intelligence field. Avery et al. proposed a framework based on TD for testing artificial intelligence algorithms [3]. The dynamic difficulty adjustment of TD was proposed in [20]. The resource allocation algorithm for the turn-based game in [11] can also be applied to TD.

However, unlike other puzzle games such as Chess and Go, the hardness of TD and DTD is still unclear since they contain many sub-problems. This research attempts to formally prove one of the sub-problems in DTD, the problem of placing turrets on the map that maximizing the distance of opponent's path toward the exit point.

© Springer International Publishing AG 2017
M. Numao et al. (Eds.): PRICAI 2016 Workshops, LNAI 10004, pp. 19–25, 2017.
DOI: 10.1007/978-3-319-60675-0_2

Fig. 1. Desktop tower defense gameplay

This paper is organized into 5 sections. Section 2, some mathematical notations are presented. The hardness of DTD is proved in Sect. 3. The analysis and discussion are elaborated in Sect. 4. The conclusion of this research is drawn in Sect. 5.

2   Preliminaries

This section starts by defining grid graph's terminology and the Hamiltonian path on the general grid graph. Then, the hardness of many games and puzzles are reviewed.

2.1   Grid Graph Terminology

Suppose G^∞ is the infinity graph with vertex set contains all points of the Euclidean plane with integer coordinates. Any two vertices in G^∞ are connected if and only if the Euclidean distance between them is 1. Let $v = (v_x, v_y)$ be a vertex in G^∞ such that v_x and v_y are integer coordinates of v in G^∞.

For any positive integer m and n, let $R(m, n)$ be the rectangular grid graph, the grid graph whose vertex set is $V(R(m, n)) = \{v | 1 \leq v_x \leq m, 1 \leq v_y \leq n\}$.

The arbitrary grid graph G is a finite vertex-induced subgraph of G^∞. Clearly, each vertex in arbitrary grid graph has degree at most 4. In other words, the arbitrary grid graph G is the subgraph isomorphism of the rectangular grid graph $R(m, n)$.

Let $G = (V, E)$ be an undirected graph, and let $s, t \in V$ be distinct vertices of G. The Hamiltonian path problem, HamPath(G, s, t), has a solution if there exists a path from s to t that visits each node in G exactly once. In the decision version, HamPath(G, s, t), is used to determine whether there is a path from s to t that visits each node in G exactly once.

The problem of finding Hamiltonian path in the arbitrary grid graph is known to be NP-complete [10]. However, there exists linear-time algorithms for some special class of grid graphs [5, 14, 21, 23].

2.2 The Hardness of Games and Puzzles

Many classic board games were proven to be EXPTIME-complete, such as Chess [19], Go [17], Chinese checkers [12], and draughts [18]. Several metatheorems for proving the hardness of modern video games were established in [8, 22]. Some modern video games, such as Price of Persia and Doom, were proven to be PSPACE-complete. Many video games, such as Tetris and Super Mario Bros, were proven to be NP-hard as well [2, 4]. Kendal provided the survey of NP-complete puzzles in [13].

3 NP-Hardness of Desktop Tower Defense

This section starts by formally defining DTD in the term of mathematical modeling. Then, the hardness of DTD is proven.

3.1 Desktop Tower Defense Problem Definition

Let m, n, k be some positive integers. Suppose T is a rectangle grid of size $m \times n$. Without loss of generality, assume that each element in T is indexed by row-major order (i.e., the square grid's index starts from $T[1][1]$ at the upper left corner to $T[m][n]$ at the lower right corner). Each element in T is marked by 0 or 1, which indicates a path and a wall respectively. Let W be a set of positions in T such that the position (x, y) in T is marked as a wall. Namely,

$$W = \{(x, y) \mid T[\mathrm{x}][\mathrm{y}] = 1 \text{ where } x \leq m \text{ and } y \leq n\}.$$

Let $\mathbf{s} = (x_s, y_s)$ be a starting point and $\mathbf{t} = (x_t, y_t)$ be a terminating point in T, for some $1 \leq x_s, x_t \leq n$ and $1 \leq y_s, y_t \leq m$. The generalized DTD, $DtD(T, W, k, \mathbf{s}, \mathbf{t})$, is to determine where to locate k additional walls to the rectangle grid T so that they can maximize the shortest path from $\mathbf{s}$ to $\mathbf{t}$. The only restriction is the position of all k walls must not completely block the terminating point $\mathbf{t}$. Therefore, there always has at least one path from $\mathbf{s}$ to $\mathbf{t}$.

The generalized DTD can be also stated in the decision form. Let d be a shortest distance from $\mathbf{s}$ to $\mathbf{t}$. The decision version, denote as $DDtd(T, W, k, \mathbf{s}, \mathbf{t}, d)$, is to determine if there is a way to place k walls in T such that the shortest path is increased to d or more. The output is "*yes*" if there is a path of length at least d after placing k additional walls, and "*no*" otherwise.

3.2   The Hardness of Desktop Tower Defense

The NP-hardness of DTD can be shown by transforming an instance of the Hamiltonian path for an arbitrary grid graph problem to an instance of the generalized DTD. Formally, the arbitrary grid graph $G \subseteq R(m, n)$ is transformed to the rectangle grid of size $(2m - 1) \times (2n - 1)$ such that the solution of the generalized DTD yields the Hamiltonian path in the arbitrary grid graph G.

Theorem 1.   The Generalized DTD is NP-hard.

Proof.   Suppose that $G = (V, E) \subseteq R(m, n)$ be an arbitrary grid graph with $|E|$ edges and $|V|$ vertices. Given the instance of the Hamiltonian path problem *HamPath*(G, s, t), we construct the instance of DTD by first placing vertices and edges of G to the $(2m - 1) \times (2n - 1)$-rectangle grid in the way such that each vertex becomes a blank square and each edge becomes a blank square. Second, flag two squares that represent vertex s and vertex t as the starting point, **s**, and terminating point, **t**, respectively. The last step, fill the rest squares that do not flagged as a blank square with walls. For example, Fig. 2 illustrates the transformation of an arbitrary grid graph $G \subseteq R(4, 6)$ to (7×11)-rectangle grid T.

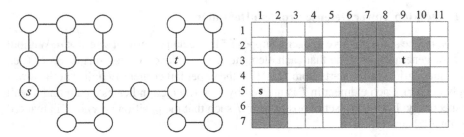

Fig. 2.   Transform grid graph to DTD

To prove that the transform process is correct, it suffices to show that the original instance of *HamPath*(G, s, t) is a *"yes"* instance if and only if the transformed *DDtd*$(T, W, k, \mathbf{s}, \mathbf{t}, d)$ instance is also a *"yes"* instance. The proof is shown in the following lemma.

Lemma 1.   *HamPath*(G, s, t) has a solution if and only if *DDtd*$(T, W, k, \mathbf{s}, \mathbf{t}, d)$, where $k = ||E| - (|V| - 1)|$ and $d = 2|V| - 1$, has a solution.

Proof.   Suppose that *HamPath*(G, s, t) has a solution. There exists a path between vertex s and vertex t that visits each vertex in G exactly once. Note that, Hamiltonian path is also a longest path from s to t. This path has length exactly $|V| - 1$ since it needs to connect $|V|$ vertices without forming a cycle. This implies that $||E| - (|V| - 1)|$ edges are not included in the Hamiltonian path of the grid graph G. Each edge in G is represented by a blank square in T. This implies that if $||E| - (|V| - 1)|$ blank squares that are not included in Hamiltonian path are filled with walls. The shortest path in T must follow the Hamiltonian path of graph G. The distance in the transformed grid can be calculated by subtracting all blank squares with the number of walls. The minimum distance from **s** to **t** is

$$d = |E| + |V| - k$$
$$= |E| + |V| - (|E| - (|V| - 1))$$
$$= 2|V| - 1,$$

since the shortest path must take all available cells. Therefore, the shortest path in $DDtd(T, W, k, \mathbf{s}, \mathbf{t}, d)$ is maximized by the longest path from s to t of G.

Suppose that $HamPath(G, s, t)$ has no solution. Then, the grid graph G does not have a Hamiltonian path between vertex s and vertex t. Therefore, there does not exist a path of length at least $|V| - 1$ that passes each node exactly once. This fact is also applied to the transformed grid T since all opponents in DTD always take the shortest path. They must not turn back to cells that have been passed. It follows that all paths from **s** to **t** in the transformed grid takes at most $2|V| - 1$ available cells. Thus, $DDtd(T, W, k, \mathbf{s}, \mathbf{t}, d)$ has no solution since it is impossible to get the minimum distance at least $2|V| - 1$ after placing $||E| - (|V| - 1)|$ walls. □

By Lemma 1, $HamPath(G, s, t) \leq_P DDtd(T, W, k, \mathbf{s}, \mathbf{t}, d)$, and the result follows. □

4 Discussion

The proof in this research only considers the problem of maximizing the minimum distance. There are many problems that are embedded in the game.

In the real gameplay, there are many types of turrets. Each of them has its own firepower, ability, range, price, and cost of upgrading. The player should determine the number of turrets to buy so that the total price is less than or equal to the given gold. Moreover, the total firepower should be large enough for intercepting enemies. This problem can be classified as 0-1 Unbounded Multiple Constraint Knapsack Problem. Gens and Levner proved that this variant of Knapsack problem is NP-complete [9].

Enemies in DTD also have distinct abilities, such as the resistance to certain types of turrets, the weakness against some types of turrets, the ability to spawn itself after getting the damage, and the ability to fly over the map. The player must plan the defense strategy for the given combination of enemies. The hardness of planning can be shown to be NP-hard using the 3-SAT framework for proving Pushing Block puzzles [6, 7].

The problem of maximizing the shortest distance does not appear in the classic TD game. Enemies in the classic TD always walk on the predetermined road. The player cannot place any obstructions on this road. Therefore, the main problem in the classic TD is to find where to place turrets such that their ranges cover the road as much as possible. This problem can be seen as the special case of the Art Gallery problem. The Art Gallery problem and its variations are also shown to be NP-hard [1, 15, 16].

5 Conclusions

This research formally proves the hardness of maximizing the shortest path problem in the Desktop Tower Defense. The hardness of DTD follows from the NP-hardness of the Hamiltonian path problem. The proof shows that the instance of $HamPath(G, s, t)$ can be transformed to the instance of $DDtd(T, W, k, \mathbf{s}, \mathbf{t}, d)$ where the number of walls is $||E| - (|V| - 1)|$ and the minimum distance is $2|V| - 1$.

There are open problems related to DTD and TD that have not been proven yet. The first problem is the resource allocation problem. It is easy to see that this problem is similar to the Knapsack problem. The second problem is the area coverage problem. This problem resembling the Art Gallery problem. The major difference between the Art Gallery problem and the area coverage in TD game is sentinels in the Art Gallery problem must cover all internal regions in the polygon. In contrast, turrets in TD only need to cover some limited area around the polygon edge.

References

1. Aggarwal, A.: The art gallery theorem: its variations, applications and algorithmic aspects. Ph.D. thesis (1984)
2. Aloupis, G., Demaine, E.D., Guo, A., Viglietta, G.: Classic nintendo games are (computationally) hard. Theor. Comput. Sci. **586**, 135–160 (2015). http://dx.doi.org/10.1016/j.tcs.2015.02.037
3. Avery, P., Togelius, J., Alistar, E., van Leeuwen, R.P.: Computational intelligence and tower defence games. In: Proceedings of the IEEE Congress on Evolutionary Computation, CEC 2011, New Orleans, LA, USA, 5–8 June, 2011, pp. 1084–1091. IEEE (2011). http://dx.doi.org/10.1109/CEC.2011.5949738
4. Breukelaar, R., Demaine, E.D., Hohenberger, S., Hoogeboom, H.J., Kosters, W.A., Liben-Nowell, D.: Tetris is hard, even to approximate. Int. J. Comput. Geometry Appl. **14**(1–2), 41–68 (2004)
5. Chen, S.D., Shen, H., Topor, R.W.: An efficient algorithm for constructing hamiltonian paths in meshes. Parallel Comput. **28**(9), 1293–1305 (2002)
6. Demaine, E.D., Demaine, M.L., Hoffmann, M., O'Rourke, J.: Pushing blocks is hard. Comput. Geom. **26**(1), 21–36 (2003). doi:10.1016/S0925-7721(02)00170-0
7. Demaine, E.D., Demaine, M.L., O'Rourke, J.: Pushpush and push-1 are nphard in 2d. In: Proceedings of the 12th Canadian Conference on Computational Geometry, Fredericton, New Brunswick, Canada, 16–19 August 2000 (2000). http://www.cccg.ca/proceedings/2000/26.ps.gz
8. Forišek, M.: Computational complexity of two-dimensional platform games. In: Boldi, P., Gargano, L. (eds.) FUN 2010. LNCS, vol. 6099, pp. 214–227. Springer, Heidelberg (2010). doi:10.1007/978-3-642-13122-6_22
9. Gens, G., Levner, E.: Complexity of approximation algorithms for combinatorial problems: a survey. SIGACT News **12**(3), 52–65 (1980). http://doi.acm.org/10.1145/1008861.1008867
10. Itai, A., Papadimitriou, C.H., Szwarcfiter, J.L.: Hamilton paths in grid graphs. SIAM J. Comput. **11**(4), 676–686 (1982)
11. Johnson, R.W., Melich, M.E., Michalewicz, Z., Schmidt, M.: Coevolutionary optimization of fuzzy logic intelligence for strategic decision support. IEEE Trans. Evol. Comput. **9**(6), 682–694 (2005)

12. Kasai, T., Adachi, A., Iwata, S.: Classes of pebble games and complete problems. In: Austing, R.H., Conti, D.M., Engel, G.L. (eds.) Proceedings 1978 ACM Annual Conference, Washington, DC, USA, 4–6 December 1978, vol. II, pp. 914–918. ACM (1978). http://doi.acm.org/10.1145/800178.810161

13. Kendall, G., Parkes, A.J., Spoerer, K.: A survey of np-complete puzzles. ICGA J. **31**(1), 13–34 (2008)

14. Keshavarz-Kohjerdi, F., Bagheri, A.: Hamiltonian paths in some classes of grid graphs. J. Appl. Math. 2012 (2012)

15. Lee, D.T., Lin, A.K.: Computational complexity of art gallery problems. IEEE Trans. Inform. Theory **32**(2), 276–282 (1986). http://dx.doi.org/10.1109/TIT.1986.1057165

16. O'Rourke, J.: Art Gallery Theorems and Algorithms. Oxford University Press Inc., New York (1987)

17. Robson, J.M.: The complexity of go. In: IFIP Congress. pp. 413–417 (1983)

18. Robson, J.M.: N by N checkers is exptime complete. SIAM J. Comput. **13**(2), 252–267 (1984). http://dx.doi.org/10.1137/0213018

19. Shannon, C.E.: Programming a Computer for Playing Chess. In: Levy, D. (ed.) Computer chess compendium, pp. 2–13. Springer, New York (1988)

20. Sutoyo, R., Winata, D., Oliviani, K., Supriyadi, D.M.: Dynamic difficulty adjustment in tower defence. In: Procedia Computer Science, pp. 435–444 (2015)

21. Umans, C., Lenhart, W.: Hamiltonian cycles in solid grid graphs. In: FOCS, pp. 496–505. IEEE Computer Society (1997)

22. Viglietta, G.: Gaming is a hard job, but someone has to do it! Theory Comput. Syst. **54**(4), 595–621 (2014). http://dx.doi.org/10.1007/s00224-013-9497-5

23. Zamfirescu, C., Zamfirescu, T.: Hamiltonian properties of grid graphs. SIAM J. Discrete Math. **5**(4), 564–570 (1992)

A Regression-Based SVD Parallelization Using Overlapping Folds for Textual Data

Uraiwan Buatoom[1,3(✉)], Thanaruk Theeramunkong[1,2],
and Waree Kongprawechnon[1]

[1] School of Information, Computer, and Communication Technology (ICT),
Sirindhorn International Institute of Technology,
Thammasat University, Bangkok, Pathum Thani, Thailand
uraiwan.buatoom@student.siit.tu.ac.th,
uraiwanb31@gmail.com,
{thanaruk,waree}@siit.tu.ac.th
[2] Associate Fellow, The Royal Society of Thailand, Bangkok, Thailand
[3] Faculty of Science and Arts, Burapha University, Chanthaburi, Thailand

Abstract. One of the most difficult issues in text mining is high dimensionality caused by a large number of features (keywords). While various multivariate analyses, such as PCA and SVD (in information retrieval, called LSI), are developed to solve this curse of high dimensionality, they are computationally costly. This paper investigates a regression-based reconstruction method that enables parallelization of PCA/SVD by decomposing a document-term matrix into a set of sub-matrices with consideration of overlapped terms, and then to re-assemble using regression technique. To evaluate our method, we utilize two text datasets in the UCI Machine Learning Repository, called "Bag of Words" and "Reuter 50 50". To measure the closeness between two documents, cosine similarity is applied while the accuracy is measured in the form of rank order mismatch. Finally, the result shows that, the matrices decomposition and re-assembly can preserve the quality of relation/representation.

Keywords: Decomposition · Re-assembly · Sub-matrix · SVD · Regression · LSI

1 Introduction

One of the most important challenges in text mining is how to handle large-scale textual datasets that usually hold characteristics of sparseness and high dimensionality with irrelevant, redundant, and noisy features, resulting in low performance but high computational cost. Towards the solution, recent works have proposed a number of data reduction [1] and data transformation methods, including feature selection [2], feature extraction [3], and dimensionality reduction [4].

Even now, it is well-known that Singular Value Decomposition (SVD) is a popular tool for feature extraction and dimensionality reduction, as for an analysis of multi-variate data. SVD can be applied to a lot of fields that share the purpose of reducing the number of features in the dataset by selecting a few singular values that may preserve the original spectrum of the values. In [5], Wall et al. showed SVD can detect and

© Springer International Publishing AG 2017
M. Numao et al. (Eds.): PRICAI 2016 Workshops, LNAI 10004, pp. 26–37, 2017.
DOI: 10.1007/978-3-319-60675-0_3

extract small signals from noisy data. However, as a side effect, it is well-known that data reduction triggers a lower performance. Although in the past there have been a large number of works on how to reduce the dimension, most of them still suffered from high computational cost [6]. In [7], Gao and Zhang reported that SVD becomes inefficient when the dataset is large.

With a specific task, such as PCA on image data where spatial dependency are essential factors in correlation calculation, Segmented-PCA (S-PCA) [11] and Folded-PCA (F-PCA) [12] were introduced and were shown to improve time complexity in feature extraction and data reduction by segmenting a data record into equal-size partition by choosing and accumulating portions in the main diagonal of the original covariance matrix. However, S-PCA and F-PCA support for image datasets [8, 9], which every feature is not related all together like text datasets.

As an alternative to dimensionality reduction, partitioning the complete feature set into a number of overlapping subgroups (proxy matrices) and performing regression on each subgroup, can help us improve the accuracy. As an example of this approach, Janire Carlos et al. [10] proposed a method to solve the problem of segmentation and to combine a complete dataset from several sub-group data. They used the regression model to combine the subset data by using the remain point in the original data to predict remain output for making the new large amount of life cycle assessment data and then reduce the time consumption.

In this paper, we propose a method that is a combination of dimensionality reduction and overlapping subgroup regression. The method firstly decomposes a large matrix into a number of smaller overlapping matrices (i.e., sub-group data), and then applies SVD to reduce the dimensions before performing regression to combine the results of such sub-group data with the consideration of the correlation between words from SVD process. While performing SVD on the large matrix triggers high computational cost, whereas applying SVD on sub-group data can be done much faster. The complexity of the proposed method is proved to be lower than the original SVD.

To evaluate our method, the cosine similarities calculated from the original matrix and those calculated from the overlapping subgroup regression are compared. Besides the absolute values, we also consider the rank order mismatches between these two types of similarities. The result showed that our method has low similarity difference and rank order mismatches. Moreover, the complexity of the proposed method is analyzed and compared with the original SVD.

The remainder of the paper is organized as follows. Section 2 describes the mathematical formulation for motivation, including singular value decompose (SVD), linear regression (LR), text segmentation, and document-term matrix. In Sect. 3, the proposed method and simulation model is illustrated. Section 4 presents experiment settings and performance measures, i.e., cosine similarity and rank order mismatch. In Sect. 5, the experimental result and error analysis are discussed. Finally, a conclusion is given in Sect. 6.

2 Motivation

This section starts singular value decomposition, regression and document representation in the form of document-term matrix.

2.1   Singular Value Decomposition and Complexity Analysis

For data compression method, the Singular Value Decomposition (SVD) is a factorization of a real or complex matrix. It is the generalization of the Eigen-decomposition of a positive semi-definite normal matrix to any $m \times n$ matrix via an extension of polar decomposition. Let X denote an $m \times n$ matrix of real-valued data (m is the number of documents and n *is the number of keywords*), where $m \leq n$ and r is the best initial good rank then $1 \leq r \leq m$. The equation of SVD of X can be represented as follows.

$$X = U \times \Sigma \times V^T \tag{1}$$

Here, U is a column-orthonormal $m \times r$ matrix, Σ is a diagonal $r \times r$ matrix with the singular values. Each element at the diagonal $r \times r$ matrix represents the eigenvalues λ_I are sorted in descending order. V is a column-orthonormal $r \times n$ matrix with r is the rank of the matrix X [11] expressed as:

$$r \in [1, \min(m,n)] \tag{2}$$

The left-singular vectors of X represents the eigenvectors of XX^T, The right-singular vectors of X also represents the eigenvectors of X^TX, and the non-zero singular values of X represents the square roots of the non-zero eigenvalues of both X^TX and XX^T [12].

In a study of SVD, the results depict that the SVD can detect and remain the characterize structure of a new matrix from decompose, although the alternate position of columns. Finally, the performed result of this experiment is shown below as:

Table 1. The comparation of a new matrix from decompose with alternate column position

X						Σ				V^T				
	1	2	3	5	7		21.9414	0	0		−0.3540	−0.4305	−0.5069	−0.3822	−0.5351
$X1 =$	4	5	6	5	7	$\Sigma1 =$	0	5.0571	0	V^T1	−0.4703	−0.3847	−0.2992	0.4276	0.5987
	7	8	9	5	7		0	0	0.0000		0.7632	−0.5953	−0.1679	0.1868	−0.0004
	5	7	1	2	3		21.9414	0	0		−0.3822	−0.5351	−0.3540	−0.4305	−0.5069
$X2 =$	5	7	4	5	6	$\Sigma2 =$	0	5.0571	0	V^T2	0.4276	0.5987	−0.4703	−0.3847	−0.2992
	5	7	7	8	9		0	0	0.0000		−0.8099	0.5745	0.0339	−0.0953	0.0615
	7	1	2	3	5		21.9414	0	0		−0.5351	−0.3540	−0.4305	−0.5069	−0.3822
$X3 =$	7	4	5	6	5	$\Sigma3 =$	0	5.0571	0	V^T3	0.5987	−0.4703	−0.3847	−0.2992	0.4276
	7	7	8	9	5		0	0	0.0000		−0.5793	−0.2022	−0.1306	0.3327	0.7041

From the real datasets, the number of real datasets in the documents (m) is smaller than the number of keywords (n). In-fact, for this research in SVD is focused only on the test matrix V which represents the relation between keywords and same like that in a study of synonymy and polysemy [5]. The SVD control methodology is to detect the type of words. V is denoted as Eigen value of keyword. The synonymy of word is defined as a strongly correlated with the same Eigen-key term

2.2   Linear Regression, Complexity Analysis and Error Estimation

The numeric values (continuous values) can be applied to the mathematics linear regression to analyze weights for combining attributes. The prediction uses the attributes, which have a relation together to transform the data from one side to another side by calculating the new values from weights [10]. The mathematic expression in this methodology represents the variable y is a model which related with independent variable x so we can define the linear regression as [13]:

$$y = w_0 + w_1 x \tag{3}$$

Here, we assume that y is a constant, and $w_0 \, and \, w_1$ are regression coefficients (weights). The regression coefficients can be expressed as:

$$\vec{w} = \left(A^T A\right)^{-1} A^T \vec{y} \tag{4}$$

To evaluate the result, it is possible to apply regressions for matching. The measurement of error data between original and predictive data can be expressed as:

$$Percent \, of \, Eror = \frac{|Measured \, Value - Actual \, Value|}{Actual \, Value} \times 100\% \tag{5}$$

2.3   Document-Term Matrix Construction

To process textual data is necessary to translate a document into a computable form. The common form is a document-term matrix, where each column is a unigram keyword feature and each row corresponds to a document, represented by a vector of terms in the document, called document vector. While there are several possibilities in encoding (weighting) terms in the document vector, some popular weighting methods are TF, TFIDF [14], and binary weighting [9], such representations are needed in text classification, text clustering, and text summarization. As the most popular weighting applied in the field of information retrieval and text mining, term frequency–inverse document frequency (TFIDF) is used for representing the importance of a word in this work.

$$TFIDF(t) = TF(t, d) \times IDF(t) \tag{6}$$

The term frequency (for short, TF) shows how many times the term t appears in the document d. The higher term frequency the term has, the more the term contributes to the document in terms of semantics. The inverse document frequency (for short, IDF) is added into the weighting (Eq. 1) since there is a trend that a term appearing in many documents has less contribution to the semantics than a term that occurs in a few documents. Even there are several alternative forms of IDF, one of the most popular equations is as follows.

$$IDF(t) = \log\left(1 + \frac{N}{n_t}\right) \tag{7}$$

where N represents the total number of documents and n_t is the number of documents that include the term t.

3 Proposed Method and Simulation Model

The design proposed methodology and simulation model is classified as follows:

3.1 Overlapping Matrix Decomposition

To reduce the dimensionality, firstly we decompose a document-term matrix into a set of sub-matrices. The model represents M rows and N columns. Here, o represents the number of columns which is overlap between sub-matrices by $o > M$, as shown in Fig. 1. and a represent the non-overlapping columns. *Let* $A \in \mathbb{R}^{m \times n}$ *is a real matrix, then* $\text{Split}(A) = [A_1, A_2, A_3, \ldots A_P]$ where $A_1 \cap A_2, A_2 \cap A_3, \ldots, A_{P-1} \cap A_P \neq \emptyset$

The function for calculating the amount of segments is as follows:

$$\begin{aligned}
N &= (2o + a) + (P - 1)(a + o) \\
&= 2o + a + Pa - a + Po - N = o + P(a + o) \\
P &= \frac{N - o}{a + o}
\end{aligned} \tag{8}$$

From Fig. 1 $a = n - 2o$ then $o = \frac{(n-a)}{2}$ So $P = \frac{N-o}{n-2o+o} = \frac{N-o}{n-o}$
If represent r is percentage for overlap area between 2 sub-group, then

$$P = \frac{N - (r*n)}{n - (r*n)} = \frac{N - (r*n)}{n*(1 - r)} \tag{9}$$

The width of range for segments equal as:

$$n = 2o + a \tag{10}$$

Where
P   is the number of partitions of sub-matrices.
o   is the number of columns that is an overlapping region of sub-matrix.
a   is the number of columns that is a non-overlapping region of sub-matrix

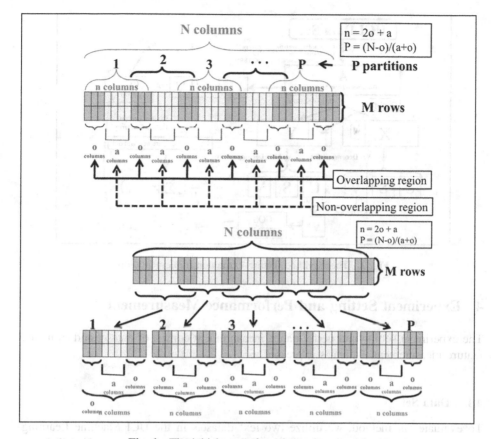

Fig. 1. The initial range for sub-matrices model

3.2 Splitting Matrix and Individual Dimensionality Reduction

In this work, we propose a method to reduce computational cost of SVD calculation on a large-scale matrix by first splitting the matrix into a number of overlapping subgroup matrices (proxy matrices), performing SVD on each subgroup matrix and then combining the result by regression techniques. This step we propose to find the optimize bound of small subgroup matrices by testing different percentage of overlapping region under control to cut rank of choosing the small diagonal entries of Σ, which consists of a descending order (mentioned in Sect. 2.1). The difference cosine similarity is given for quality measurement of result between original methodology and re-assembly methodology. The complete model of modified feature extraction as shown in Fig. 2.

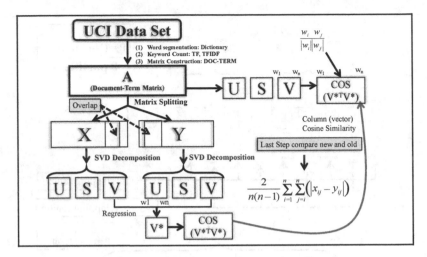

Fig. 2. The modified feature extraction implementation

4 Experiment Setting and Performance Measurement

The experiment setting and performance measurement consist of dataset and stimulate feature measurement, which is expressed as:

4.1 Data Set

To evaluate our method, we utilize two text datasets in the UCI Machine Learning Repository [15, 16], namely "Bag of Words" (later, BOW) and "Reuter 50 50" (later, C50). Due to the sake of computational complexity, for the former dataset, we select the "NIPS full papers" subgroup, which is composed of 1,500 documents with 12,419 distinct words, and approximately 19,000,000 words in total. The latter dataset, i.e., C50, contains 2,500 documents. In the experiments, we select 1,500 from 2,500 documents due to the computational reason (14,284 distinct keywords, nearly 21,500,000 words in total.). For document-term representation, the TF-IDF weighing is used.

4.2 Stimulate Feature Measurement Similarity Between Word

The new matrices V from SVD, focus on the relation of word view. The results from 2 groups estimate the efficiency by measuring the similarity in 2 ways and it is discussed as follows:

4.2.1 Cosine Similarity
The Cosine Similarity is one of the popular measurement, which is used for measuring the similarity of vectors. This measurement proposed to measure the same composition of vectors by cosine of the angle. Garcia [17] has shown that the matrix of the new

vector coordinates, which was reduced dimensional space from SVD process for the closest of matrix, which has higher score of cosine similarity than other vectors. The estimated equation is follow as

$$\cos(w_1, w_2) = \frac{w_1.w_2}{\|w_1\|.\|w_2\|} \frac{\sum_{t_i=t_j} w_i.w_j}{\sqrt{\sum w_i^2 \sum w_j^2}} \qquad (11)$$

4.2.2 Rank Order Mismatch

The Rank order mismatch is the method to measure the order of ranking data. Pavan et al. shown to use this tool to measure the quality of order sequence of text structure and they described when data is most quality ranking then the value of rank will nearly be one [18]. The proposed methodology highlights the comparison of difference of order ranking between similarity keywords. The resulted cosine similarity measurement method is used as an input of this method. The quality of order ranking expresses rank equation as:

$$\begin{aligned} \text{Rank}(w_1, w_2) &= \frac{\text{Match}(w_1, w_2)}{\text{Match}(w_1, w_2) + \text{MisMatch}(w_1, w_2)} \\ &= \frac{\sum_{i=1}^n \sum_{j=i-1}^m \text{Match}(w_i, w_j)}{\sum_{i=1}^n \sum_{j=i-1}^m \text{Match}(w_i, w_j) + \sum_{i=1}^n \sum_{j=i-1}^m \text{MisMatch}(w_i, w_j)} \end{aligned}$$

$$(12)$$

Where, $\text{MisMatch}(w_i, w_j)$ is the number of mismatched term in batch of w_1, w_2. $\text{Match}(w_i, w_j)$ also represents the number of matching rank of both w_i and w_j. Moreover, Comparison loop shows the loop assigned remark score by considering value of between w_i and w_j In case of w_i, it is greater than w_j then remark score is +0.5. In case

	W1	W2	W3	W4	W5
W1	1	0.2	0.4	0.3	0.5
W2	0.2	1	0.7	0.1	0.3
W3	0.4	0.7	1	0.4	0.5
W4	0.3	0.1	0.4	1	0.6
W5	0.5	0.3	0.5	0.6	1

a

	W1	W2	W3	W4	W5
W1	1	0.4	0.3	0.2	0.5
W2	0.4	1	0.7	0.5	0.2
W3	0.3	0.7	1	0.4	0.3
W4	0.2	0.5	0.4	1	0.5
W5	0.5	0.2	0.3	0.5	1

b

+

	Method 1	Method2	Result of mismatch
W2w3	-0.5	+0.5	1
W2w4	-0.5	+0.5	1
W2w5	-0.5	-0.5	0
W3w4	+0.5	+0.5	0
W3w5	-0.5	-0.5	0
W4w5	-0.5	-0.5	0

c

Fig. 3. **a.** The v* matrix from method 1 **b.** The v matrix from method 2 **c.** The desired result of rank mismatch

of w_i, it is lesser than w_j then remark score is -0.5 and when the value is equal to assign remark the score is 0. Finally, the rank order mismatch model is shown below as (Fig. 3).

5 Results and Discussion

The proposed methodology is applied to regression to combine the sub-matrices. Comparing the results by different percentage of overlapping of dataset. We found that from Table 2 (row 1st and 2nd) that the error of predicted for mapping data is always less than 1% by using the both datasets. Figure 4. Shows the Comparison performance of prediction for mapping data. Most of the results shown the trend error of data stability is in lowest range between 30% and 35% for overlapping region columns.

The methodology applied statistics to measure quality of the result two main topics. Which is about similarity of relation word between original matrices and new data from re-assembly matrices and compare reduce cost of time from stimulate.

5.1 Measuring Similarity of Relation of Words

Table 1, the row 3rd and 4th elaborates the two groups of dataset for comparing the different percentage of overlapping data. These values are statistics of similarity cosine. We can infer that the performance of several datasets follows the degree of percentage of overlapping data. However, we could also infer that the different similarity of words between the lowest and highest range is not much different. So, the most of data in re-assembly matrix is closest to the original data matrix.

Table 2. The error of predict regression and different similarity cosine results (no. of rows = 1500)

Measurement method	The percentage of overlapping	15	20	25	30	35
Error regression	R-BOW	0.01370	0.00380	0.00270	0.00083	0.00079
	R-C50	0.94980	0.08430	0.05790	0.01190	0.01010
Different similarity cosine	DC-BOW	0.01640	0.01420	0.01210	0.01030	0.00910
	DC-C50	0.02240	0.01760	0.01370	0.00800	0.00650

Table 3. The result of rank order mismatch

Dataset	Match(w_i, w_j)	MisMatch(w_i, w_j)	Rank
BOW (overlapping 35%)	907,483,622,718	50,140,139,531	0.94764
C50 (overlapping 35%)	1,384,715,601,633	72,383,611,191	0.95032

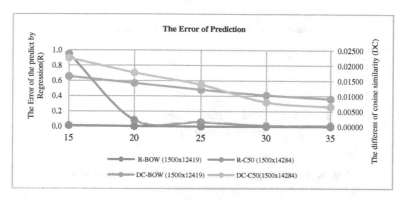

Fig. 4. The error of predict regression and different similarity cosine results

Moreover, to get the best results, we choose the best result of different similarity cosine from both mentioned two groups to test the rank order mismatch by checking every column. Table 3 uses the same dataset of Table 2. Table 3 shows the performance in the view of how much performance for ranking order mismatch of data from similarity cosine between words. The best rank is C50 data set at 0.95032. The result approach 1 shows that the meaning order rank of data between two groups is nearly the same ranking. Finally, the results show that the rank is nearly 1.

5.2 Measuring Reduce Cost of Simulation

Table 4 highlights the three group type of data which is reformed by the different sub-matrices data. Table 4. also compare the result of the average SVD runtimes from five times. That shows the 3 sub-matrices group have the lowest time. At Fig. 5 also shows the trend of data from three groups have low running time followed by the amount of sub-groups data. Therefore, we can conclude that, if we reform data to many sub-groups then it ensures that the running time will decrease and the results will follow to calculate the BIG-O table.

Although, Fig. 5 depicts the graph in 3 groups (P), which is segmented in original matrix to 3 sub-matrices by overlapping10% region, has sharply dropped of running time follow the size of data. Moreover, Table 5 compares the performance of BIG-O. This also shows the performance of Covariance matrix and Eigen Problem follow the higher number of partitions and the less of percentage of overlapping. However, in mapping matrices by linear-regression had overhead, but it still has the lower cost than old-SVD.

Table 4. The performance of running time

Time running SVD (secs)	The size of Data (M × M) Segmented Sub-group (P) overlapping 10%					
	100	1,000	5,000	10,000	15,000	20,000
1 groups (P)	4.89×10^{-3}	7.32×10^{-1}	7.35×10	5.29×10^2	1.95×10^3	4.73×10^3
2 groups (P)	3.37×10^{-3}	2.74×10^{-1}	2.71×10	1.87×10^2	6.37×10^2	1.47×10^3
3 groups (P)	2.56×10^{-3}	1.46×10^{-1}	1.23×10	8.92×10	1.87×10^2	6.54×10^2

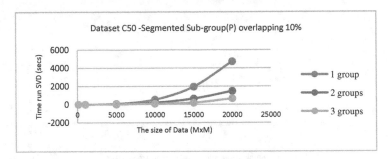

Fig. 5. Comparing performance of SVD by running time

The proposed BIG-O between 2 methods is as follows:

Table 5. The BIG-O of method between SVD and Modified sub-matrices + SVD

Process	SVD (A)	Sub-matrices + SVD (B)	Ratio (A/B)
Covariance matrix size	$O(MN^2)$	$O(Pmn^2)$	r^2P
Eigen problem	$O(N^3)$	$O(Pn^3)$	$\left(\frac{r^2}{1-r}\right)P$
Mapping by linear regression	–	$O((P-1)(r*n))$	–

where, M, m: The number of samples

N, n: The dimensions of constructed matrix for computation

P: The number of times to mapping data

o: The number of overlapping region (from Sect. 3.1 $o = \frac{(n-a)}{2}$)

r: The percentage of overlapping region

6 Conclusion

Effect of data segment by using dimension reduction methodology, we can reduce the cost of time, then we can also decrease the size of matrix by re-assemble to sub-matrix. The proposed methodology, the segment sub-matrix, under the condition shows that, the set of column size of example data is bigger than the row data. The column size data represents the number of set example data for finding the weights of coefficient of linear regression. The linear regression can be used as a tool for combining matrix back to the same size of original matrix. The experiment results also claim that with the measurement of the different similarity cosine of word correlation, which is gained from SVD, between original matrix and sub-matrix. To check the better performance, the rank order mismatch is used, which has high value matching i.e. nearly equal to one.

This paper shows that, we still have a chance to re-build a new matrix which has less overlapping region under less timing computation and still preserve the quality of matrix. However, to reform the back size regression in this methodology still has the high BIG-O, so, we can also focus on the cost of predictive data for combining the less than linear regression. Finally, we can improve a number of group to segment, which could be studied and experiment in the future.

Acknowledgement. This work is financially funded and supported by Sirindhorn International Institute of Technology, Thammasat University and Burapha University.

References

1. Chen, Y.H., Ting-Chia, L.: Dimension reduction techniques for accessing Chinese readability. In: Machine Learning and Cybernetics ICMLC (2014)
2. Ketui, N., Theeramunkong, T.: Effect of weighting factors and unit-selection factors on text summarization. In: Pham, D.-N., Park, S.-B. (eds.) PRICAI 2014. LNCS (LNAI), vol. 8862, pp. 891–897. Springer, Cham (2014). doi:10.1007/978-3-319-13560-1_75
3. He, Q., Ding, X.: Sparse representation based on local time-frequency template matching for bearing transient fault feature extraction. J. Sound Vib. **370**, 424–443 (2016)
4. Bharti, K.K., Singh, P.K.: A three-stage unsupervised dimension reduction method for text clustering. J. Comput. Sci. **5**(2), 156–169 (2014)
5. Wall, M.E., Rechtsteiner, A., Rocha, L.M.: Singular value decomposition and principal component analysis. In: Berrar, D.P., Dubitzky, W., Granzow, M. (eds.) A Practical Approach to Microarray Data Analysis, pp. 91–109. Springer, Boston (2003)
6. Jun, S., Park, S.-S., Jang, D.-S.: Document clustering method using dimension reduction and support vector clustering to overcome sparseness. Expert Syst. Appl. **41**(7), 3204–3212 (2014)
7. Gao, J., Zhang, J.: Clustered SVD strategies in latent semantic indexing. Inf. Process. Manage. **41**(5), 1051–1063 (2005)
8. Zabalza, J., et al.: Novel Folded-PCA for improved feature extraction and data reduction with hyperspectral imaging and SAR in remote sensing. ISPRS J. Photogrammetry Remote Sens. **93**, 112–122 (2005)
9. Xiuping, J., Richards, J.A.: Segmented principal components transformation for efficient hyperspectral remote-sensing image display and classification. IEEE Trans. Geosci. Remote Sens. **37**(1), 538–542 (1999)
10. Pascual-González, J., et al.: Combined use of MILP and multi-linear regression to simplify LCA studies. Comput. Chem. Eng. **82**, 34–43 (2015)
11. Qiao, H.: New SVD based initialization strategy for non-negative matrix factorization. Pattern Recogn. Lett. **63**, 71–77 (2015)
12. Shlens, J.: A tutorial on principal component analysis (2003)
13. Theeramunkong, T.: Introduction to concepts and techniques in data mining and application to text mining (2012)
14. Kittiphattanabawon, N., Theeramunkong, T., Nantajeewarawat, E.: News relation discovery based on association rule mining with combining factors. IEICE Trans. **94**, 404–415 (2011)
15. Lichman, M.: UCI Machine Learning Repository (2013). http://archive.ics.uci.edu/ml
16. ZhiLiu, UCI Machine Learning Repository (2011). https://archive.ics.uci.edu/ml/datasets/Reuter_50_50
17. Garcia, D.E.: Latent Semantic Indexing (LSI) A Fast Track Tutorial (2006)
18. Pavan Kumar, P., Agarwal, A., Bhagvati, C.: A structure based approach for mathematical expression retrieval. In: 6th International Workshop Multi-disciplinary Trends in Artificial Intelligence, MIWAI, Ho Chi Minh City, Vietnam (2012)

Virtual Reality System with Smartphone Application for Height Exposure

Suppanut Nateeraitaiwa and Narit Hnoohom[(✉)]

Image, Information and Intelligence Laboratory, Department of Computer Engineering,
Faculty of Engineering, Mahidol University, Nakorn Pathom, Thailand
Suppanut.n@hotmail.com, narit.hno@mahidol.ac.th

Abstract. One of the treatment methods for phobias is behavioral therapy by creating patient's fear environment for the patient to confront that situation. Virtual reality is one of the most interesting technologies for creating three-dimensional virtual environments. The virtual reality technology makes users feel like they are immersed in a virtual world. To accomplish creating fear environment, this paper presented a virtual reality system with a smartphone application for height exposure. The virtual reality system is simple, which consists of software and hardware. Users can use this system easily in their own home. With an evaluation on 20 participants, the impact of user involvement on realism and fear of heights was explored. Paired samples t-test showed that the user's influence on the height of the building was significant. Moreover, the user's influence on the realism and fear of heights when the sound is activated were significant.

Keywords: Height exposure · Virtual reality · Smartphone application · Virtual reality glasses · Remote controller

1 Introduction

Virtual reality technology is creating three-dimensional virtual environments by computers. The technology makes users feel like they are immersed in a virtual world and that they can control their avatar by action in the real world [1]. Since the year 1990, virtual reality technology has become popular. It has been developed and used in various fields, particularly in entertainment, as well as a tool to help treat some diseases [2].

A phobia is a type of anxiety disorder, which has an impact on a patients' quality of life. Patients with phobias have excessive and irrational fear of a situation or a particular object. The fear happens often and patients cannot stop their fear. Patients will avoid behavior to confront that situation, making their life miserable [3]. The examples of phobias include acrophobia (fear of heights) [4], cynophobia (fear of dogs) [5], arachnophobia (fear of spiders) [6], and claustrophobia (fear of confined spaces) [7].

One of the treatment methods for phobias is behavioral therapy by creating a patient's fear environment for the patient to confront that situation. Virtual reality technology can create a virtual environment that makes users feel like they immersed in a virtual world. We can simulate a virtual environment that displays a patient's fear situation.

© Springer International Publishing AG 2017
M. Numao et al. (Eds.): PRICAI 2016 Workshops, LNAI 10004, pp. 38–50, 2017.
DOI: 10.1007/978-3-319-60675-0_4

Virtual reality exposure therapy is provided to patients immersed in a computer generated virtual environment, either through the use of a head-mounted display (HMD) device or entry into computer-assisted virtual environments (CAVEs) where images are present all around. Computer will create and control that virtual environment to show the situation that patient fears. You can see that using virtual reality to create patients' fear situation is help patients can directly confront patients' fear situation without risks of doing the same in real life [8].

Hodges et al. [9] described a pilot study that uses a virtual reality technology for treating acrophobia. The result showed subject had to act in accordance with the situation show in the virtual world that subject facing and, subject has difference level of anxiety in difference situations. In terms of emotional processing theory, the fear structure of subjects showed clearly that the subjects' responses to facing situation and, the subjects' responses clearly that anxiety of subjects is reduced. The result support that the fear structure are changed when treated.

Haworth et al. [10] were proposed a virtual reality system for treating phobias: Acrophobia and Arachnophobia. This virtual reality exposure therapy (VRET) system is a low-cost, and readily available software and hardware components. The system using Kinect to track patient's body. However, the system must always be connected to internet for communicating between the clinician and the patients.

Schafer et al. [11] have presented a virtual reality technology to create virtual environment and player avatar for exploring the effective treatment of acrophobia. The selected 42 subjects divided into two groups. First group uses the system with avatar and second group uses the system only. The result showed significant differences in score of two groups.

This paper is organized as follows: first, a general introduction to the appearance of acrophobia and treatment guidelines using virtual reality technologies are briefly described. The hardware and software requirements for this research are given in Sect. 2. Next, Sect. 3 considers the method to develop the smartphone application. The experimental settings and results are shown in Sect. 4. Finally, this research is concluded in Sect. 5.

2 System

The system is composed of hardware and software components that work together to simulate a virtual environment. This system is a head-mounted display set up with a smartphone to display a virtual environment. We simulate it and use a wireless remote controller for control of the avatar in the virtual world.

2.1 Hardware Selection

The related hardware of the research is illustrated in Fig. 1, contains main 4 devices: display monitor, gyroscope sensor, virtual reality glasses, and remote controller. The following sections present a brief summary of each device.

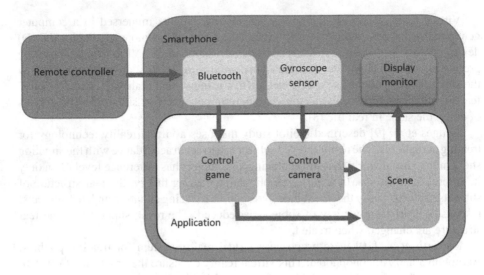

Fig. 1. Block diagram of system process

Display monitor: We use a smartphone which runs an android operating system for display. These days a smartphone is an important device. Almost everyone has their own smartphone and android is a popular smartphone operating system. On September 29, 2015, Google announced that there are currently 1.4 billion active android devices worldwide [12]. For developers, the Android software development kit (SDK) is an open source software. Android has the cardboard SDK for our integrated development environment (IDE) to easily adapt an existing 3D application for virtual reality (VR).

Gyroscope sensor: We use a gyroscope sensor in the smartphone. The gyroscope measures the rate of rotation in radian per second around a device's x, y, and z axes. Rotation is positive in the counter-clockwise direction; that is, an observer looking from some positive location on the x, y or z axes at a device positioned on the origin would report positive rotation if the device appeared to be rotating counter-clockwise. This is the standard mathematical definition of positive rotation and is not the same as the definition for roll that is used by the orientation sensor. This research uses the value from the gyroscope as input to control the camera view in the application.

Virtual reality glasses: A pair of virtual reality glasses is a simple head-mounted display that can allow a smartphone to display two images, one for the left eye and one for the right. Virtual reality glasses contain two polarized lenses for adjusting the focus of the eyes. The Virtual reality glasses used in this research are a pair of 3D Shinecon Glasses, as shown in Fig. 2.

Remote controller: The basic controller devices are well compatible with an android smartphone. Communication via Bluetooth. Device is also simple and easy to use. Figure 3 shows a universal wireless remote controller for controlling the movement of an avatar in the virtual world.

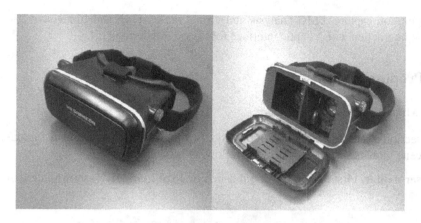

Fig. 2. 3D VR Shinecon glasses

Fig. 3. Universal wireless remote controller

2.2 Software Selection

Unity 5.3.2 is used as an engine for developing our application. Unity can import Cardboard SDK for support to develop a virtual reality application and can build the application for use with many platforms of smartphones. This SDK makes it easy to:

- Build an existing 3D application for mobile application by using a display of the mobile application that does not deviate from the Unity 3D application.
- Adapt an existing Unity 3D application to a virtual reality application.
- We can control the camera viewing to be compatible with head tracking.

Unity is a game engine for developing virtual reality game applications that we can download from Play Store, for example RollerCoaster, Crazy Swing, and Zombie Run developed by FiBRUM, and VR Roller Coaster, VR Cave Flythrough, and VR Volcano Flythrough developed by frag. And there are some papers that use Unity for developing applications as well, for example a serious game with virtual reality for travel training for those with Autism Spectrum Disorder [13], design of a video game for rehabilitation using motion capture, EMG analysis and virtual reality [14], and Immersive VR for natural interaction with a haptic interface for Shape Rendering [15].

3   Propose Methods

3.1   Model Selection

This section discusses the features of the models, which are used in the scene for making our scene more realistic. We select the following models.

- **User avatar.** First Person Lover published by ISBIT GAME[1,2], shown in Fig. 4.

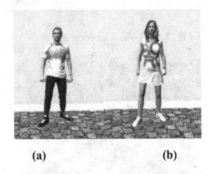

(a)   (b)

Fig. 4. (a) Male avatar model, (b) Female avatar model

- **Teleport Effect.** KY Magic Effect published by Kakky[3], shown in Fig. 5.

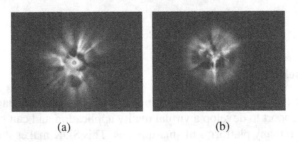

(a)   (b)

Fig. 5. (a) Green teleport effect, (b) Orange teleport effect (Color figure online)

---

[1] https://www.assetstore.unity3d.com/en/#!/content/40848.
[2] https://www.assetstore.unity3d.com/en/#!/content/41056.
[3] https://www.assetstore.unity3d.com/en/#!/content/21927.

- **Buildings for height exposure.** Medieval Buildings published by 7th Dimension[4], Block Building Pack published by CGY (Yemelyan K.)[5], Radio Tower - Low Poly published by VR[6], shown in Fig. 6.

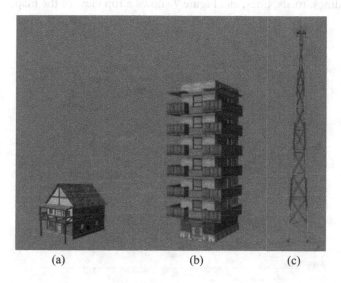

(a)   (b)   (c)

Fig. 6. (a) Building Level 1 balcony & roof, (b) Building Level 2 seventh floor springboard, (c) Building Level 3 peak tower

3.2 Map Design

The objective of this research is to use a virtual reality application to simulate a virtual environment for users to confront several levels of height situations. The scene in the application simulates the user avatar in a virtual city. Users can control his/her avatar to move in a designated area. In the designated area, there are 3 buildings that users can move upward to the top of in order to confront different levels of heights. Users can move up a building by teleporting which shows a scene like a green effect. When users move into the green effect, the user's avatar will teleport to the top of this building at a fixed position. And users can teleport down by moving into an orange effect, then users will teleport to the start position. The three buildings are different structures as follows:

- Level 1 balcony & roof: This building is a two-story house that has a balcony, as shown in Fig. 6(a).
- Level 2 seventh floor springboard: Fig. 6(b) shows a high building, which has 7 floors. On top of the building there is an open terrace.

---

[4] https://www.assetstore.unity3d.com/en/#!/content/34770.
[5] https://www.assetstore.unity3d.com/en/#!/content/13925.
[6] https://www.assetstore.unity3d.com/en/#!/content/2299.

- Level 3 peak tower: This building is a radio tower that is very high, as shown in Fig. 6(c).
- Other objects in the scene are intended to make the scene more realistic, for example other buildings, roads, trees, etc. Figure 7 shows a top view of the map.

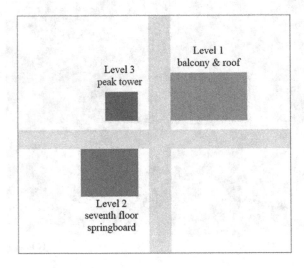

Fig. 7. Top view of map

3.3 Avatar Design

User avatars are created by the object character controller from Unity. The Avatar has a capsule shape that height of 6 Unity units (unit of Unity) and a radius of 1.5 Unity units. On top of the capsule there is a main camera for display of output. The radius of the capsule uses this value to always keep the main camera on a capsule collider. The character controller has a script for the control of movement of the user's avatar, so jumping, animation and rotation follow the camera view, illustrated in Fig. 4.

3.4 Teleport Design

To teleport is to go from one place to another place. If we want to translate the avatar position to somewhere that the user cannot get to using basic movement, such as to the top of the tower, we use teleport to solve this problem. This application has 2 types of teleport. First is the green teleport and the second is the orange teleport. Users can move into the green teleport in the scene, shown in Fig. 5(a), then the user's avatar will translate the position to the top of that building. If the user wants to bring the avatar down, the user can teleport down by moving into the orange teleport, shown in Fig. 5 (b), and then the user avatar will come back to the start position. The script of the teleport has 3 inputs, which are x, y, and z for translating the user's avatar by combination with input x, y and z.

4   Results and Discussions

4.1   Results

We obtained a virtual reality system for height exposure. The system is a head-mounted display that put on a user's head. Users can control his/her character via the wireless remote controller, as seen in Fig. 8. The application is a virtual reality application that has 2 paths of display, one for each eye. When the application starts, the user's character will appear at the start point, illustrated in Fig. 9, and the menu will appear on the display for training the control of the character. Users can control character movement in the map, and users can use the teleport for moving up the buildings to start the height exposure.

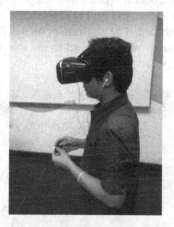

Fig. 8.   The system

Fig. 9.   Display output in virtual reality model

Level 1 balcony & roof: When user teleport upward, shown in Fig. 10, their user avatar will stay on the balcony of this house. That balcony has a railing and is not so high and then users can continue to teleport upward to the house's roof from the balcony. That roof does not have a railing like the balcony, but is still not so high.

Fig. 10. Scene of balcony and roof

Level 2 seventh floor springboard: When a user teleport upward, the avatar will stay at the fringe of the terrace, as shown in Fig. 11. The terrace has a platform outstretched from the terrace like a springboard. The platform is on the 7th floor with a small area, without a railing. When the user moves upward to the top of this building, the sound effect will be changed to a wind sound.

Fig. 11. Scene of seventh floor springboard

Level 3 insular: When a user teleport upward, the avatar will stay at the top of the radio tower with a very small area and the height is very high, as shown in Fig. 12. When users move upward to the top of the radio tower, the sound effect will be changed to a wind sound.

Fig. 12. Scene of top of tower

4.2   Discussion

In order to evaluate the performance of the proposed system, we conducted a questionnaire survey on 20 participants. The questionnaire consists of two parts. The first part contains questions relating to demographic data (gender, age, virtual reality experience, fear of heights). In the second part, included 16 questions which the view of participants reflected on the factors that might affecting the realism of the virtual environment and fear of heights.

The total number of participants is 20 persons, 13 men and 7 women, aged 15–40. Among the 20 persons, 5 persons have virtual reality experience and 2 persons have a fear of heights as displayed in the Table 1.

Table 1. Participants

Characteristics		Frequency	Percentage
Gender	Male	13	65
	Female	7	35
Age	15–20	3	15
	20–25	12	60
	25–30	4	20
	35–40	1	5
Virtual reality experience	Always	1	5
	Sometimes	4	20
	Never	15	75
Fear of heights	Yes	2	10
	No	18	90

Data for the analysis are the scores received from the questionnaires. The scores were set from 1–4 as follows: 1 = few, 2 = average, 3 = much and 4 = very much. After the analysis by paired samples t-test with reliability at 95% (alpha = 0.05) to compare the scores of every question, the results show that when users work in the high building, the fear of heights becomes higher depending on the height of the building with significance (comparing building level 1 with level 2, $t = 4.682$, $p = 0.000$), (comparing building level 2 with level 3, $t = 2.333$, $p = 0.031$). In comparison between the score of realism and fear of heights when the sound is activated and deactivated, the score when the sound is activated is higher than with the deactivated sound with significance (when realism $t = -9.200$, $p = 0.000$), (fear of height $t = 6.097$, $p = 0.000$). When users can see their avatar model, the realism score is more than when they are unable to see the avatar with significance ($t = 2.449$, $p = 0.024$). Additionally, when the avatar model suitable with user to represent gender identity (male or female), the users will have an increased score of fear with significance ($t = 2.990$, $p = 0.008$), while the scores of fear when users are able to see their avatar or unable to see it do not differ from each other ($p = 0.385$). Also, there is no difference in the realism of the virtual environment scores when the avatar model and the user are suitable or unsuitable ($p = 0.104$).

In conclusion, the height of the building, the sounds and the use of the avatar model which matches the user affects the fear of heights score rate the same as the use of sound and the user being able to see their avatar model affects the score of realism of the virtual environment. While playing, the moment that users can see the avatar model or cannot see it does not impact the fear of heights the same as choosing the model of their avatar to suitable with the user does not affect the realism of the virtual environment score. The means of each topic are displayed in the Table 2 and comparison by paired samples t-test is displayed in the Table 3.

Table 2. Descriptives

Score of (1-4)	Frequency	Mean	SD
Realism of virtual environment	20	2.30	0.733
Fear of heights at building level 1	20	1.20	0.523
Fear of heights at building level 2	20	1.95	0.686
Fear of heights at building level 3	20	2.30	0.865
Realism when sound activated	20	2.65	0.745
Realism when sound deactivated	20	1.60	0.681
Fear of heights when sound activated	20	2.60	0.754
Fear of heights when sound deactivated	20	1.85	0.745
Realism when user's avatar activated	20	2.65	0.875
Realism when user's avatar deactivated	20	2.05	0.826
Fear of heights when user's avatar activated	20	2.40	0.883
Fear of heights when user's avatar deactivated	20	2.20	0.768
Realism when using avatar model suitable with user	20	2.15	0.933
Realism when using avatar model unsuitable with user	20	1.95	0.826
Fear of heights when using avatar model suitable with user	20	2.25	0.851
Fear of heights when using avatar model unsuitable with user	20	1.85	0.875

Table 3. Paired samples t-test

Pairs	t	df	p
Fear of heights at building level 1 - Fear of heights at building level 2	−4.682	19	0.000
Fear of heights at building level 2 - Fear of heights at building level 3	−2.333	19	0.031
Realism when sound activated - Realism when sound deactivated	9.200	19	0.000
Fear of heights when sound activated - Fear of heights when sound deactivated	6.097	19	0.000
Realism when user's avatar activated - Realism when user's avatar deactivated	2.449	19	0.024
Fear of heights when user's avatar activated - Fear of heights when user's avatar deactivated	0.890	19	0.385
Realism when using avatar model suitable with user - Realism when using avatar model unsuitable with user	1.710	19	0.104
Fear of heights when using avatar model suitable with user - Fear of heights when using avatar model unsuitable with user	2.990	19	0.008

Comparing the score of realism of the virtual environment of the application to 2 at 90% of reliability (alpha = 0.1), the received score is more than 2 with significance (p = 0.083). To conclude, the score of realism of the virtual environment of the application is ranked highly and at the highest level as shown in the Table 4.

Table 4. One sample t-test

Score of (1–4)	t	df	p
Realism of virtual environment	1.831	19	0.083

5   Conclusion

We have created an application to be used for exposure to heights that runs on smartphones, as currently all types of smartphones are already widely used. Users just buy a few extra accessories to help create a virtual reality world. These accessories are virtual reality glasses and a wireless remote controller. The price is affordable. Users can use this application for confrontation with height situations at his/her own home, or anywhere. The hardware system is a standalone head-mounted display that does not connect with any wires. The software system is used to simulate and display a virtual environment via smartphones. In a virtual environment, users can confront 3 levels of height situations. Each level has a different height. Users can start with a low level of height and increase the level of height when users can pass the previous level. Application scenes are developed by Unity engine. The design is deployment object models in scenes that are easy-to-use and provide realism. In the future work, we are planning to use a new controller that can be used to perform more interactions with the virtual environment, and to make the application capable of connecting to the internet in order to contact psychiatrists to monitor users when using the application and give suggestions from the psychiatrist in the application scenes.

Acknowledgements. This work is supported by the Department of Computer Engineering, Faculty of Engineering, Mahidol University.

References

1. What is virtual reality? - virtual reality. http://www.vrs.org.uk/virtual-reality/what-is-virtual-reality.html
2. Virtual reality, technology of the future. Episode 1. https://blog.eduzones.com/darkfairytale/35
3. Ruangtrakool, S.: Textbook of Psychiatry. Ruenkaew Printing, Bangkok (1999)
4. Costa, J.P., Robb, J., Nacke, L.E.: Physiological acrophobia evaluation through in vivo exposure in a VR CAVE. In: IEEE Games Media Entertainment, pp. 1–4. IEEE, Toronto (2014)
5. Benavides, C.: Virtual reality in the treatment of cynophobia. In: 10th Computing Colombian Conference (10CCC), pp. 499–503. IEEE, Bogota (2015)

6. Cavrag, M., Lariviere, G., Cretu, A.-M., Bouchard, S.: Interaction with virtual spiders for eliciting disgust in the treatment of phobias. In: IEEE International Symposium on Haptic, Audio and Visual Environments and Games (HAVE) Proceedings, pp. 29–34. IEEE, Richardson (2014)
7. Bruce, M., Regenbrecht, H.: A virtual reality claustrophobia therapy system - implementation and test. In: IEEE Virtual Reality Conference, pp. 179–182. IEEE, Lafayette (2009)
8. Tull, M.: How virtual reality exposure therapy (VRET) treats PTSD, 18 April 2016. https://www.verywell.com/virtual-reality-exposure-therapy-vret-2797340
9. Hodges, L.F., Kooper, R., Meyer, T.C., Rothbaum, B.O., Opdyke, D., de Graaff, J.J., Williford, J.S., North, M.M.: Virtual environments for treating the fear of heights. Computer **28**, 27–34 (1995)
10. Haworth, M.B., Baljko, M., Faloutsos, P.: PhoVR: a virtual reality system to treat phobias. In: Proceedings of the 11th ACM SIGGRAPH International Conference on Virtual-Reality Continuum and its Applications in Industry - VRCAI 2012, pp. 171–174. ACM (2012)
11. Schafer, P., Koller, M., Diemer, J., Meixner, G.: Development and evaluation of a virtual reality-system with integrated tracking of extremities under the aspect of Acrophobia. In: SAI Intelligent Systems Conference (IntelliSys), pp. 408–417. IEEE, London (2015)
12. Callaham, J.: Google says there are now 1.4 billion active Android devices worldwide. http://www.androidcentral.com/google-says-there-are-now-14-billion-active-android-devices-worldwide
13. Bernardes, M., Barros, F., Simoes, M., Castelo-Branco, M.: A serious game with virtual reality for travel training with autism spectrum disorder. In: International Conference on Virtual Rehabilitation (ICVR), pp. 127–128. IEEE, Valencia (2015)
14. Rincon, A.L., Yamasaki, H., Shimoda, S.: Design of a video game for rehabilitation using motion capture, EMG analysis and virtual reality. In: International Conference on Electronics, Communications and Computers (CONIELECOMP). pp. 198–204. IEEE, Cholula (2016)
15. Covarrubias, M., Bordegoni, M.: Immersive VR for natural interaction with a haptic interface for shape rendering. In: IEEE 1st International Forum on Research and Technologies for Society and Industry Leveraging a better tomorrow (RTSI), pp. 82–89. IEEE, Turin (2015)

Classification of Diabetic Retinopathy Stages Using Image Segmentation and an Artificial Neural Network

Narit Hnoohom[✉] and Ratikanlaya Tanthuwapathom

Image, Information and Intelligence Laboratory, Department of Computer Engineering,
Faculty of Engineering, Mahidol University, Nakorn Pathom, Thailand
narit.hno@mahidol.ac.th, t.ratikanlaya@gmail.com

Abstract. Diabetic retinopathy, which can lead to blindness, has been found in
22% of diabetic patients in the latest survey. Therefore, diabetic patients should
have an eye examination at least once a year. However, it has been found that
currently there is a problematic lack of specialists in ophthalmology. Detection
and treatment of diabetic retinopathy are thus delayed. The idea to create a classification system of diabetic retinopathy stages to facilitate the making of preliminary decisions by ophthalmologists is introduced. This paper presents the classification of diabetic retinopathy stages using image segmentation and an artificial
neural network. This proposed method applies local thresholding to separate the
foreground region from the background region so that the optic disc and exudate
regions are able to be identified more clearly. The experiment was carried out
with 100 fundus images from the Institute of Medical Research and Technology
Assessment database. The prediction model had an accuracy rate of up to 96%.

Keywords: Diabetic retinopathy · Exudates · Fundus image · Artificial neural
network

1 Introduction

Due to a genetic disease characteristic, 371 million people around the world currently
have diabetes mellitus, and 500 million diabetic patients are expected by 2030. In Thailand, the number amounts to approximately 3.5 million patients with diabetes mellitus.
In addition, a recent survey found that 22% of people with diabetes mellitus have diabetic
retinopathy, which can lead to blindness. Furthermore, the risk of loss of eyesight among
people with diabetes mellitus is at least 20% higher than ordinary people. Although
diabetic patients may experience the same pathologies as ordinary people, e.g. cataracts,
glaucoma and optic nerve terminal inflammation, these pathologies are encountered at
younger ages with greater frequency, severity and treatment complexity than in other
people, even though all of these symptoms can be treated with the same methods used
for non-diabetics [2, 13]. However, there is one critical condition that is only found in
diabetic patients, and that is diabetic retinopathy, which usually occurs in patients who
have had diabetes mellitus for a long time. According to previous findings, patients who
have had diabetes mellitus for less than 10 years are at 7% greater risk for diabetic
retinopathy. However, the risk increases to 63% for patients with diabetes mellitus for

© Springer International Publishing AG 2017
M. Numao et al. (Eds.): PRICAI 2016 Workshops, LNAI 10004, pp. 51–62, 2017.
DOI: 10.1007/978-3-319-60675-0_5

more than 15 years and patients with good glucose control can still have diabetic retinopathy in the long term or at an older age.

The treatment of patients with eye problems requires an ophthalmologist with advanced instruments. Although there are currently around 700 ophthalmologists in Thailand, this number is insufficient for the increasing number of diabetic patients. Thus, diabetic retinopathy screening is considered an important process in helping patients. In order to treat this group of patients in time, all patients with diabetes mellitus should have their eyes examined at least once every year to thoroughly check for symptoms of diabetic retinopathy.

This article focuses on the development of detection and classification methods that can be utilized for diabetic retinopathy screening. In this area, several methods have been developed and the area is still being explored. These methods can be divided into two categories: blood vessel detection and exudate detection.

The blood vessel detection method is used to detect and classify the amount of blood vessels in the retinal image. K. Verma et al. [1] presented the classification of moderate and non-proliferative phases of diabetic retinopathy (NPDR) by finding the amount of blood vessels and blood in the retinal images. Blood vessels in the retinal image can be used to classify the differences between the vascular area and background. The principle means of the segmentation in this research was thresholding and the use of the dot-blot hemorrhage microaneurysm (MA) as the exudate indicator. This method employs the use of a retinal image database with a total of 65 fundus images consisting of 30 images showing normal stages, 23 images showing moderate stages and 12 images showing severe stages from a structured analysis of the retina. The result of the system test was a correct classification rate of 93%.

Exudate detection is used to detect and classify the amount of exudates in the retinal image. Z. Ahmad et al. [2] presented the development of this detection and classification system with segmentation. The DR classification process depends on the amount of exudates. In the retinal images, the system is able to help ophthalmologists perform early screening of patients with diabetes mellitus. There are two processes of exudates detection; rough and fine exudates segmentation. Rough segmentation is performed by using the morphology operation and column-wise neighborhoods operation, while a good classification should be done by using the morphological reconstruction. This method employs the use of a retinal image database with a total of 239 fundus images consisting of 110 images showing normal stages, 63 images showing mild stages, 36 images showing moderate stages and 30 images showing severe stages from Sungai Buloh Hospital in Malaysia together with more appropriate retinal feature extractions. This classification method can also provide a simple and fast basis for analysis. The result of the system test was 60% correct classification.

S. Sreng et al. [3] proposed early detection of diabetic retinopathy using the green channel and red channel, exudate extraction with the image binarization, region of interest (ROI) based segmentation and morphological reconstruction. This method employs the use of a retinal image database with a total of 100 fundus images from Bhumibol Adulyadej Hospital. The result of the system test was 91% correct classification.

M. Eman Shahin et al. [4] presented the method for detection of exudates in retina images using a morphological and candy method. Following microaneurysms detection using histogram equalization, morphological and candy edge detector, this method employs the use of a retinal image database with a total of 340 fundus images consisting of 89 images showing normal stages and 251 images showing abnormal. The result of the system test was 92% correct classification by ANN.

Du Ning et al. [5] presented the method of using computers for monitoring the diabetic retinopathy stage using a color fundus images process. The techniques of morphological processing and method of texture analysis are applied to the retinal images for the feature examination, for example, vascular areas, exudates, sharpness of the identical areas, and features that will be added to the support vector machine (SVM). This method employs the use of a retinal image database with a total of 52 fundus images consisting of 10 images showing normal stages, 35 images showing NPDR stages and 7 images showing proliferative phases of diabetic retinopathy (PDR) stages from the standard diabetic retinopathy database. This research showed the correct classification of diabetic retinopathy at 93% and detected hard exudates in color fundus images of the human retina.

C. Jayakumari et al. [6] presented the method of exudates detection using the following process: histogram equalization then segmentation of the image using a contextual clustering algorithm. The selected sets of features are the standard deviation of the intensity, mean of intensity, area, edge strength and learning system by the echo stage neural network (ESNN). The images are classified class into two categories, which are normal and abnormal. A total of 50 images are used to find the exudates, out of which 35 images are used to train the ESSN and the remaining 15 images are used to test the neural network. This article confirms 93% sensitivity and 100% specificity in terms of exudates based classification.

This paper presents a classification method for diabetic retinopathy screening by using image segmentation and an artificial neural network to classify retina images of diabetic retinopathy at the normal, moderate, and severe stages. The proposed method applies a local thresholding to separate the foreground region exudates from the background region. This can identify the optic disc and exudate regions in fundus images more clearly. Detection and extraction of the optic disc are obtained using a morphological operation. In order to demonstrate the reliability of the method, the proposed method is compared with the results of the examination by ophthalmologists. The proposed method also applies related principles in order to accelerate the treatment of patients.

The remainder of the paper is organised as follows. First, a general introduction to the detection method of diabetic retinopathy is briefly described. In Sect. 2, a description of the proposed method with the processes of classification of diabetic retinopathy stages is addressed. Experiment results and discussions are included in Sect. 3. Finally, the conclusions are presented in Sect. 4.

2   Proposed Method

The main objective of the proposed method is to classify the diabetic retinopathy stages. The proposed method is shown in Fig. 1, and contains the following main five processes: (1) Image acquisition; (2) Pre-processing; (3) Exudate extraction; (4) Features extraction; and (5) the Artificial neural network. The following sections present a brief summary of each step.

Fig. 1. Overview of the proposed method

2.1   Image Acquisition

The first phase of the classification of diabetic retinopathy stages is the image acquisition. The fundus images were collected from the Institute of Medical Research and Technology Assessment (MRTA) database, which contains 100 images. The database consists of both normal retinal images and abnormal retinal images. The set of normal retinal images contains 18 images and the set of abnormal retinal images contains 82 images, in which 47 images are moderate retinal images and 35 images are severe retinal images. The original fundus images, which are a size of 3872 × 2592 pixels in the Portable Network Graphics (PNG) file format, were captured by the fundus photographing optical system.

2.2   Pre-processing

During the pre-processing stage, the fundus image data is prepared for use in the classification of diabetic retinopathy stages.

2.2.1   Image Resize
Each fundus image from MRTA database is usually in different sizes. The standard size of the input fundus images used in this proposed method is defined as 512 × 512 pixels. This can reduce computation time necessary for the detection of the exudate regions. Figure 2 shows the result image after resizing.

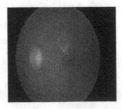

Fig. 2. Image after resizing

2.2.2 Edge Removal

To remove the edge of the resized image, we delete the edges to trim the brightness at the edges of the image using region of interest processing. The results of reducing the brightness of the image area is not a problem for segmentation in the next step. The result after removing edge is shown in Fig. 3.

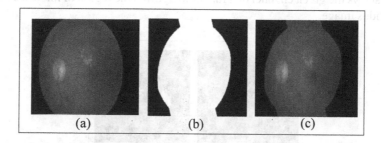

Fig. 3. (a) Resized image, (b) Region of interest, (c) Removed edge image

2.2.3 Color Model Selection

In this step, we split the removed edge image into three parts: red channel, green channel and blue channel. When the channels are separated, the findings were as follows.

In Fig. 4, it can be seen that the red channel is the channel with the most visibly intense light, but the differences of intensity in each component is relatively low making it hard to distinguish the composition. The green channel is the channel that can best show the differences in light intensity of each component of the images with the optic disc and exudates being the most intense. The blue channel is found to have the lowest color intensity and cannot be used to help in the feature differences analysis.

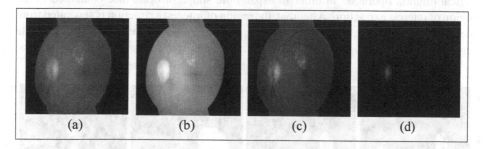

Fig. 4. (a) Removed edge image, (b) Red channel, (c) Green channel, (d) Blue channel

2.3 Exudate Extraction

Exudate extraction is composed of two stages: the region of interest extraction and the optic disc removal.

2.3.1 Region of Interest Extraction

The region of interest extraction consists of the optic disc and exudate regions in the green channel of fundus images. In order to extract the region of interest, this proposed method applies a local thresholding to separate the foreground region from the background region so as to be able to identify the optic disc and exudate regions more clearly. Figure 5 shows the green channel of a fundus image and the region of interest extraction in the fundus image.

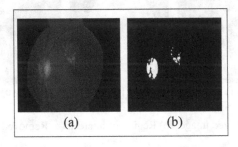

Fig. 5. (a) Green channel image, (b) Optic disc and exudate regions extracted using local thresholding

2.3.2 Optic Disc Removal

This section consists of two steps: the optic disc extraction and the optic disc removal. The optic disc region is obtained by removing the exudate regions in the result of the region of interest extraction using a morphological opening operation, which includes erosion and dilation methods in processing [7–9].

In the optic disc removal, the completed optic disc region is created by the morphological dilation operation in order to expand the shape of optic disc contained in the region of interest extraction. After that, the region of interest extraction is separated from the complete optic disc region in order to remove the optic disc region, and the exudate regions are obtained. Figure 6 shows the various stages of the optic disc removal process.

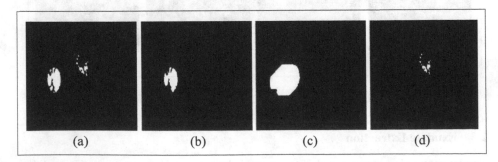

Fig. 6. (a) Region of interest extraction, (b) Optic disc extraction, (c) Completed optic disc obtained using dilation, (d) Exudate region extracted using morphological operation

2.3.3   Connected Area Analysis

Finding the exudates by using adjoining regions is a very good method. Therefore, this paper adopts this method for the study. The researchers were able to classify the exudates by considering neighboring pixels to help decide whether or not the exudates were present. On the other hand, if the neighboring pixels do not meet the specified criteria, the image will be considered to be contaminated by the noise within it, as shown in Fig. 7.

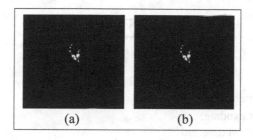

(a)   (b)

Fig. 7. (a) Output of exudate extraction with green channel, (b) Output of exudate extraction with gray scale

2.4   Features Extraction

Feature extraction is used to find the special features of the exudate extraction in retinal images. This paper used eleven features for classification of diabetic retinopathy stages:

- Standard deviation (SD) of the intensity, which is white pixels, is used to quantify the amount of variation of exudate regions in the fundus image. The SD formula is defined as follows:

$$SD_{intensity} = \sqrt{\frac{1}{N-1} \sum_{i=1}^{N} |A_i - \mu|^2} \tag{1}$$

where

$SD_{intensity}$ is the standard deviation of intensity,
A is the area of exudate,
N is the number of exudates and
μ is the mean of intensity in the fundus image

- Mean of intensity, is the average value of the gray scale image. Mean intensity can be determined by Eq. (2).

$$\mu_{intensity} = \sqrt{\frac{1}{N-1} \sum_{i=1}^{N} A_i} \tag{2}$$

Where

$\mu_{intensity}$ is the mean of intensity,
A is the area of exudate and
N is the number of exudates

- Sum of intensity, is the total of intensity, which can be determined by Eq. (3).

$$S_{intensity} = \sum_{i=1}^{N} A_i \qquad (3)$$

Where

$S_{intensity}$ is the sum intensity,
i is number of exudate,
A is the area of exudate and
N is the number of exudates

- Edge strength, is measured as the average of the edge values in the perimeter of the region. The edge values were obtained after the application of a Prewitt operator. The edge strength formula is defined as follows:

$$ES = \frac{1}{MN} \sum_{i=1}^{M} \sum_{j=1}^{N} E_i \qquad (4)$$

Where

ES is the edge strength,
MN is the total pixel of exudate,
E is the edge value of exudate and
i,j is the position of pixel

- Compactness, is another measure of circularity where A and P are the area and perimeter of the candidate region. Compactness can be determined by Eq. (5).

$$C = \frac{P^2}{4\pi A} \qquad (5)$$

Where

C is exudate compactness,
A is the area of exudate and
N is the exudate perimeter

- Area, is the area of exudates. Area can be determined by Eq. (6).

$$A = \sum_{i=1}^{N} \sum_{j=1}^{N} B[i,j] \qquad (6)$$

Where

A is the exudate area,
B is the exudate and
i,j is the position of pixel

- SD of hue, saturation, and value (HSV), is applied to quantify the amount of variation of the exudate regions in the H, S and V.
- SD of the green channel, which is white pixels, is used to quantify the amount of variation of exudate regions in the green channel.
- Perimeter is the distance around the exudate regions.
- Major axis length is the longest diameter of the exudate regions.
- Minor axis length is the shortest diameter of the exudate regions.

2.5 Classification

This paper describes exudate classification by an artificial neural network. Training is a process of using samples to develop the artificial neural network utilizing the types of input with the correct answers [12]. In this process, the group of samples with known output is repeatedly sent to the network to train the network system. In the proposed method, the training process was carried out until there were differences between input and output, and the pattern for the training set value was acceptable. Several methods are available for the network training. The back-propagation method is commonly used and is capable of working successfully in two steps. In the first step, input is sent forward through the network to produce output. In the second step, the difference between real output and expected output generates an error signal that returns the value over the network to improve the weight of the input.

Research on diabetic retinopathy has been given a lot of attention and there are numerous articles that have proposed many methods. Thus, it is difficult to compare our algorithm with the other reported work in the literature. The paper from Jayakumari [6] was chosen for comparison because there were findings on the exudates detection for classification of diabetic retinopathy stages and some features that were used identically.

3   Results and Discussions

This paper aimed to evaluate the performance by accuracy, precision and recall. The algorithm used in the classification of diabetes retinopathy was employed for experimentation with WEKA 3.7.1 for comparison. System learning with a series of 100 retinal images, training and testing was conducted with the WEKA system by a multilayer perceptron classification. In this paper, the system learning value was set at a learning rate of 0.3, a momentum of 0.2, a training time of 500, and a 10-fold cross-validation for the two sets to be tested. The Jayakumari features set comprised the SD of intensity, mean of intensity, sum of intensity, edge strength, compactness and area. The proposed features set comprised the SD of HSV, SD of green channel, perimeter, major axis length and minor axis length.

The performance series features of the two sets of algorithms in the multilayer perceptron test compared the efficiency of the two features by measuring the effectiveness of information classification, precision and recall, by which accuracy was calculated by Eqs. (7), (8) and (9):

$$Accuracy = \frac{TP + TN}{TP + TN + FN + FP} \tag{7}$$

$$Recall = \frac{TP}{TP + FN} \tag{8}$$

$$Precision = \frac{TP}{TP + FP} \tag{9}$$

Where recall is the recollection of classification categories for each group.
Precision is the accuracy of the classification categories for each group.
True Positive (TP) is the prediction of p, and the actual value is p.
False Positive (FP) is the prediction of p, but the actual value is n.
True Negative (TN) is the prediction of n, and the actual value is n.
False Negative (FN) is the prediction of n, but the actual value is p.
p is the correct prediction for the model.
n is the incorrect prediction for the model.

From Table 1, the Jayakumari features set had a result for accuracy of 94%, for precision of 94.1% and for recall of 94%. The proposed features set had a result for accuracy of 96%, for precision of 96.2% and for recall of 96%. As a result, the proposed features set yielded a higher accuracy than the Jayakumari features set.

Table 1. Results of qualifying features set for the work

Feature sets	Accuracy	Precision	Recall
Jayakumari features set SD of intensity Mean of intensity Sum of intensity Edge strength Compactness Area	94%	94.1%	94%
Proposed features set SD of HSV SD of green channel Perimeter Major axis length Minor axis length	96%	96.2%	96%

From the experiments, we found that the SD of HSV feature and the SD of green channel feature were better than the SD of intensity feature. That brings good results because in the process of segmentation in the green channel there is more detail than when using the gray scale. The perimeter feature is more suitable for this paper than the area feature, because the shape of the exudates has a variety of forms. Therefore, the perimeter feature had better results than the area feature. When we know the major axis length and the minor axis length, we can know the shape of exudates.

Furthermore, it was found that the proposed exudates screening will work well when the input fusdus image is clear with the optic disc and exudates clearly seen. The presentation of this research can be compared to the research conducted by Sreng [3] with the elimination of the optic disc even if the optic disc is not circular.

4 Conclusions

This paper presented the classification of diabetic retinopathy stages using image segmentation and an artificial neural network. The data sets are separated into three groups that are the set of normal retinal images and the set of abnormal retinal images, in which are moderate retinal images and severe retinal images. The groups were classified by exudates detection, and if detected, the exudate in the retinal image is categorized as a moderate or severe class. Nevertheless, the proposed method has a limitation in the optic disc extraction, which is that the optic disc image must be clearly seen. In future research work, this can be extended in order to classify the mild stages of diabetic retinopathy.

Acknowledgment. This project is supported by Department of Computer Engineering, Faculty of Engineering, Mahidol University. The authors would like to thank Institute of Medical Research and Technology Assessment for the database.

References

1. Verma, K., Deep, P., Ramakrishnan, A.G.: Detection and classification of diabetic retinopathy using retinal images. In: India Conference (INDICON), pp. 1–6. IEEE Press, India (2011)
2. Ahmad Zikri, R., Hadzli, H., Syed, F.: A proposed diabetic retinopathy classification algorithm with statistical inference of exudates detection. In: 2013 International Conference, Electrical, Electronics and System Engineering (ICEESE), pp. 80–95. IEEE Press, Malaysia (2013)
3. Sreng, S., Maneerat, N., Isarakorn, D., Pasaya, B., Takada, J., Panjaphongse, R., Varakulsiripunth, R.: Automatic exudate extraction for early detection of diabetic retinopathy. In: International Conference on Information Technology and Electrical Engineering (ICITEE), pp. 31–35. IEEE Press, Thailand (2013)
4. Shahin, E.M., Taha, T.E., Al-Nuaimy, W., El Rabaie, S., Zahran, O.F., El-Samie, F.E.A.: Automated detection of diabetic retinopathy in blurred digital fundus images. In: 8th International Computer Engineering Conference (ICENCO), pp. 20–25, UK (2013)

5. Ning, D., Yafen, L.: Automated identification of diabetic retinopathy stages using support vector machine. In: 32nd Chinese on Control Conference (CCC), pp. 3882–3886. IEEE Press, China (2012)
6. Jayakumari, C., Maruthi, R.: Detection of hard exudates in color fundus images of the human retina. Procedia Eng. **30**(2012), 297–302 (2012)
7. Saravanan, V., Venkatalakshmi, B., Farhana, S.M.N.: Design and development of pervasive classifier for diabetic retinopathy. In: 2013 IEEE Conference on Information & Communication Technologies (ICT), pp. 231–235. IEEE Press, India (2013)
8. Zeljkovic, V., Bojic, M., Tameze, C., Valev, V.: Classification algorithm of retina images of diabetic patients based on exudates detection. In: 2012 International Conference on High Performance Computing and Simulation (HPCS), pp. 167–173. IEEE Press, New York (2012)
9. Dhiravidachelvi, E., Rajamani, V.: Computerized detection of optic disc in diabetic retinal images using background subtraction model. In: 2014 International Conference on Circuit, Power and Computing Technologies (ICCPCT), pp. 1217–1222. IEEE Press, India (2014)
10. Usman Akram, M., Shehzad, K., Shoab Khan, A.: Identification and classification of microaneurysms for early detection of diabetic retinopathy. Pattern Recogn. **46**(1), 107–116 (2013)
11. María, G., Clara Sánchez, I., María López, I., Daniel, A., Roberto, H.: Neural network based detection of hard exudates in retinal images. Comput. Meth. Programs Biomed. **93**(1), 9–19 (2009)
12. Anitha, J., Selvathi, D., Hemanth, D.J.: Neural computing based abnormality detection in retinal optical images. In: IACC 2009. IEEE International on Advance Computing Conference, pp. 630–635. IEEE Press, India (2009)
13. Jensen, J.R.: Calculate Confusion Matrices. 20 Retrieved December 2015. Web site: http://www.exelisvis.com/docs/CalculatingConfusionMatrices.html

AIED 2016: Workshop on Artificial Intelligence for Educational Applications

Building a Semantic Ontology for Virtual Peers in Narrative-Based Environments

Ethel Chua Joy Ong[(⊠)], Danielle Grace Consignado,
Sabrina Jane Ong, and Zhayne Chong Soriano

Center for Language Technologies, De La Salle University, Manila, Philippines
ethel.ong@delasalle.ph

Abstract. Narrative-based environments utilize various forms of knowledge to provide an interactive space for the learner and the virtual agent to collaborate in accomplishing the learning goals. In this paper, we present the design of a semantic ontology that provides the necessary domain-based conceptual knowledge to allow a virtual peer to engage in storytelling as a form of exchange with the learner. We then show how the ontology was utilized to support the virtual peer in performing its tasks, which include generating interactive stories that teach about appropriate social behavior, and engaging in a text-based dialogue with the learner.

Keywords: Semantic ontology · Virtual peer · Story generation · Dialogue generation · Commonsense knowledge

1 Introduction

Narrative-based environments have utilized virtual agents to serve various roles, such as playmates or learning companion [1, 2], teachable peer [3], facilitator, and tutor [4]. A virtual playmate has a similar developmental age and understands the world in similar ways as the child-user [5]. Its goal is to stimulate the child's learning by collaborating on shared tasks and even competing for producing quality work output. A teachable peer, on the other hand, reverses the concept of peer learning by having the child-user take on the tutor role, thus advocating learning by teaching. A virtual facilitator shows the student around the virtual learning environment, alerts him/her to what is new or relevant, and assists in navigating through the lesson [6]. As a tutor, the virtual agent monitors the performance and gives appropriate feedback during all pedagogical stages (acquisition, application and assessment).

For virtual agents to be effective in carrying out their roles, they must be given domain knowledge about the lessons to be covered, pedagogical knowledge about teaching strategies and remediation, linguistics knowledge to provide appropriate feedback and responses, skills for intervening at the proper time during a learning session, and even personality to motivate the student to begin, continue and complete the required learning activities. For narrative-based environments, the agents must also be given conceptual knowledge about concepts on everyday things that the learners are familiar with, and which they can use when engaging learners in storytelling as a form of pedagogical strategy or intervention.

© Springer International Publishing AG 2017
M. Numao et al. (Eds.): PRICAI 2016 Workshops, LNAI 10004, pp. 65–76, 2017.
DOI: 10.1007/978-3-319-60675-0_6

In this paper, we present the design of a semantic ontology that provides the necessary domain-based conceptual knowledge to allow a virtual peer to engage in storytelling as a form of exchange with the learner. Interaction with virtual peers can be more natural and engaging if the peer is made to exhibit human-like behavior. In our study, we advocate storytelling as the means to achieve this natural interaction. People naturally engage in storytelling as a form of communication to narrate about daily life experiences, share beliefs and exchange information. Furthermore, storytelling remains extensively used in modern classrooms to enhance the learning experience of children [7].

The use of semantic ontology to provide the necessary domain-based conceptual knowledge to a story generation system has been explored in [8–11]. The semantic ontology contains commonsense knowledge about the world that enable people to understand each other. Giving a similar body of knowledge to software agents will enable computer systems that can understand and generate text in natural language to be developed.

The paper also details how the semantic ontology was utilized to support two virtual peers in performing their tasks. Ellie, presented in Sect. 2, is a virtual peer that teaches appropriate social behavior to children with autism. Carla, presented in Sect. 3, is a conversational peer who engages the learner in a text-based dialogue. The paper ends with a discussion of the lessons learned from our study.

2 The Semantic Ontology of Ellie, the Virtual Social Peer

Ellie is a virtual peer designed for an interactive storytelling environment that teaches children with autism about proper social behavior. The semantic ontology built for Ellie contains assertions in the form of binary relations about concepts, events and their relationships that are relevant to the themes of the storytelling system. Currently, these themes revolve around teaching social values on *waiting for turn*, *greetings*, *tidying up*, and *sharing*.

A primary source of commonsense knowledge for the semantic ontology of Ellie is ConceptNet [12]. It is a publicly available, large semantic graph containing nodes of concepts that represent words or phrases in natural language, and edges that represent the relationships between two concepts. Two reasons, however, necessitate the need to build a separate ontology for Ellie.

First, ConceptNet has been populated by crowdsourcing data through the Open Mind Common Sense project [13]. As such, assertions that are not appropriate for the target audience, specifically children diagnosed with mild autism, abound. Only concepts relevant to the story themes were extracted for use by Ellie.

Second, the interactive stories narrated by Ellie adopted the Social Story structure developed by Carol Gray [14]. In a Social Story™, a situation is described in terms of relevant social cues, perspectives and common responses in order to share accurate social information in a patient and reassuring manner. The rationale behind every desirable action, social norms and reaction to common situations is also continuously explained clearly and explicitly as facts to prepare the learners for social interactions.

2.1 Conceptual Relations

Given the previous requirements, the conceptual relations in the semantic ontology of Ellie were categorized into two – *classic* relations and *social* relations. *Classic* relations are those used to describe common concepts and events. They include the following relations adapted from ConceptNet: *usedFor*, *locatedAt*, *isFor*, and *can*. Table 1 provides a brief description and example assertions for these.

Table 1. Classic semantic relations to describe concepts and events.

Relations	Description	Example assertions
usedFor	Indicates the purpose of an object	usedFor(ball, catch) usedFor(swing, play) usedFor(book, read)
locatedAt	Specifies where an item may be situated in the virtual world	locatedAt(swing, playground) locatedAt(teachers, school) locatedAt(book, school)
isFor	Describes an activity that may take place in the specified location	isFor(playground, everyone to share and have fun)
can	States what the specified story character can do	can(children, say "Hello!") can(teachers, smile politely)

Classic relations enable Ellie to generate story text that contain descriptions of objects as well as to provide possible expectations in a new situation. Consider the sample assertions in Table 1. Ellie can use the *usedFor* relations to describe the purpose of an object in the virtual story world. For example, from the *usedFor(book, read)* assertion, Ellie can generate *"You can <u>read</u> a <u>book</u>."*. The *locatedAt* relation can be used to prepare the child for what he/she may find in a given virtual world location, e.g., *"Sometimes you meet new people in places like the <u>school</u>. You might encounter <u>teachers</u> or other children."*

Social relations, on the other hand, contain positive assertions that embody a "social sense" to describe commonsense knowledge relevant to social interactions [9]. Table 2 provides a brief description and example assertions with "social sense".

Social relations are used by Ellie as a means of preparing the learner to engage in social interactions. Consider the sample assertions in Table 2. Assuming the setting of the story is still the school, Ellie can use the *mayFeel* and *mayThink* relations to let the learner be aware of his/her possible feelings and thoughts, and to assure that these are acceptable, e.g., *"Sometimes you may feel <u>shy</u> and think <u>about leaving</u>. That's okay."*

The combined used of *classic* and *social* relations allow Ellie to also generate story text that teaches about social behavior. For example, the *can* relations in Table 1 can lead to the generation of a story text that informs the learner on how to respond in the given situation, e.g., *"You can say 'Hello!'"*

Table 2. Semantic relations with "social sense" to describe social concepts.

Relations	Description	Example assertions
mayFeel	Relates what the user might feel at a certain event	mayFeel(school, shy) mayFeel(school, nervous) mayFeel(playground, irritated)
mayThink	Suggests what the user may think at a certain event	mayThink(school, about leaving) mayThink(school, would rather play)
sharedBy	Defines who or what can share an object	sharedBy(ball, children) sharedBy(book, classmates)
expectedTo	Conveys what the user may expect at a given location	expectedTo(playground, see other people) expectedTo(playground, see other children run around)
whenUntidy	Describes the state of an object when it is untidy	whenUntidy(book, lying on the floor) whenUntidy(toys, scattered on the ground)

2.2 Generating Story Text

Children with autism often have difficulty generalizing what they have learned in one social setting to another [15]. The availability of variances in the semantic ontology, combined with templates, allows for the generation of stories with the same theme but situated in different settings, to help prepare learners for many possible situations. For example, the presence of two *expectedTo* assertions can lead to the generation of story text that may prepare the learner to *see other people in the playground*, or to *see other children run around in the playground*. This is similar to the notion presented by the Social Story Idea Toolkit [9], which also utilizes ConceptNet [12] as a resource to aid writers of Social Stories™.

An example template that is used to generate the body of the story is shown in Listing 1. The body of a Social Story™ contains sentences that add further description to the topic sentence that was stated in the story's introduction. The use of template-based text generation technique is necessary to ensure that the resulting sentences are appropriate for the target learners, as validated by special education teachers.

A template contains tags used to define the category of data that can be used as values for the tags. Tags in angle brackets (<>) are used to retrieve story world elements, such as the current location and the name of the non-playing character that the learner, who serves a "player" role, is interacting with. Tags in square brackets ([]) are used to query the ontology.

Listing 1. A template to generate the story body.

Sometimes you meet new people in places like the <location>. You might encounter [locatedAt] or [locatedAt]. They [can] or [can]. Sometimes you might feel [mayFeel] and think [mayThink]. That's okay.

For example, if the <*location*> tag returns "*school*", then the [*locatedAt*] tag is used to query the ontology with "*locatedAt(?, school)*". Using the sample assertions in Tables 1 and 2, the resulting concept is "*teachers*".

Depending on the available assertions, a set of candidate concepts may be returned by the ontology. For example, the [*mayFeel*] tag, which queries the ontology with "*mayFeel(school, ?)*", receives the candidate concepts "*shy*" and "*nervous*". In such a situation, the story generator randomly selects from the candidate concepts to instantiate a given template.

The template in Listing 1 may lead to the generation of any one of the two sample story text shown in Listing 2, with tags replaced by underlined words and phrases.

Listing 2. Sample story text generated from the template given in Listing 1.

Sometimes you meet new people in places like the <u>school</u>. You might encounter <u>fellow classmates</u> or <u>teachers</u>. They <u>say hello</u> or <u>smile politely</u>. Sometimes you might feel <u>shy</u> and think <u>about leaving</u>. That's okay.

Sometimes you meet new people in places like the <u>store</u>. You might encounter <u>shoppers</u> or <u>staff members</u>. They can <u>say hello</u> or <u>smile politely</u>. Sometimes you might feel <u>shy</u> and think <u>that you should ignore them</u>. That's okay.

Ellie also uses predefined dialogue templates to explain the rationale behind a good decision that the learner has made in the interactive storytelling environment, a sample of which is shown in Listing 3 for the *Greetings* theme.

Listing 3. A dialogue template that is used when a good decision has been made.

Other people can be the first one to greet you. You can initiate greetings too. People appreciate it when you greet them. Let's practice greetings! You can [can]. You can also [can] or [can].

All the discussions thus far portray Ellie as having the role of a teacher who adopts a narrative format to describe social situations and teach about proper behavior. But placed in an interactive storytelling environment, Ellie occasionally switches her role to a facilitator to engage the learner in a decision-making activity. Specifically, Ellie presents options on how the learner would want the story to proceed, as shown in

Listing 4. The underlined phrases in the second type of decision-making (presentation of the story problem) are concepts derived from the ontology.

Listing 4. Decision-making points in the interactive story world.

Presentation of Tasks	What do you want to do now?
	Option 1: Look around
	Option 2: Play with the slide
Presentation of the Story Problem	When you want to <u>play with the slide</u>, others might want to <u>play with the slide</u> too. What should you do?
	Option 1: Push others away
	Option 2: Wait in line

3   The Semantic Ontology of Carla, the Conversational Peer

Carla is a virtual peer that is integrated to a learning environment for reading short stories and answering reading comprehension exercises. Carla is designed to engage the learner (children who are 8-10 years old) in a text-based dialogue in an attempt to shift the learner's negative affect to one that is positive. Intelligent Tutoring Systems such as AutoTutor [4] have been enhanced to take into consideration the learner's affect state, which has an effect on his/her learning performance. Carla's design posits that a positive affect can motivate the completion of the required learning activity, in this case, that of answering reading comprehension exercises.

3.1   Populating the Semantic Ontology

Carla uses a semantic ontology to provide the possible topics of discourse that it can use during its conversation with the learner. The commonsense knowledge comprising this ontology has been directly sourced from ConceptNet 5 [16]. Of the 24 relations in ConceptNet, only six are presently being used. These are enumerated in Table 3 with the corresponding question that can be answered by the assertions of the given relation.

Table 3. Semantic relations in the ontology of Carla. (C*n* refers to Concept *n*)

Relation	Answerable question	Sentence template
isA	What kind of a thing is it?	<C1> is a kind of <C2>
definedAs	How is it defined?	<C1> is [a/an/the] <C2>
hasProperty	What property/ies does it possess?	<C1> is <C2>
hasA	What feature/s does it have?	<C1> has <C2>
madeOf	What is it made of?	<C1> is made of <C2>
locatedAt	Where can you find it?	You can find <C1 > at <C2>

Using the questions enumerated in Table 3 as the basis, some assertions extracted from ConceptNet were replaced with more suitable relations, such as replacing *hasProperty(red, one of primary color)* with *definedAs(red, primary color)*, and *hasProperty(ice cream, make of fruit)* with *madeOf(ice cream, fruit)*. The availability of the most suitable relation for a pair of given concepts is important in order to use the correct sentence template during dialogue generation.

The current iteration of the semantic ontology of Carla consists of over 600 commonsense assertions that are age-appropriate to the target audience and that are in the domains of everyday objects and activities at home and at school, food and sports. These domains were selected to complement the reading materials.

3.2 Generating Dialogue Turns

Carla uses the assertions in the semantic ontology to form the text that comprises its dialogue turn. Specifically, the conversation begins with a statement that expresses a commonsense thought, e.g., "*A fruit is good for you.*" In this instance, the topic used to start the discourse, "*fruit*", came from the reading material which is about a young boy whose painting contains images of fruits.

An open-ended question that is related to the first statement follows, e.g., "*What else is good for you?*", as well as a list of candidate responses. The latter is derived by querying the ontology for assertions that use the same relation as the first statement, to retrieve concepts that are semantically related from the ontology. A detailed example is shown in Table 4.

Table 4. Queries to the ontology to derive the contents of Carla's dialogue.

Text	Query	Candidate assertions
Statement	?(fruit, ?)	hasProperty(fruit, good for you) → if selected by planner locatedAt(fruit, grocery store)
Options	hasProperty(?, good for you)	hasProperty(exercise, good for you) hasProperty(sleep, good for you) hasProperty(warm bath, good for you)

Continuing from the given example, the user is presented with the following options as his/her response to the question "*What else is good for you?*".

"*Exercise is good for you.*"

"*Sleep is good for you.*"

"*Warm bath is good for you.*"

In case more than three candidate assertions were retrieved from the ontology, three will be randomly chosen.

If the user selects *"Exercise is good for you."*, the ontology is again queried to find candidate assertions using *"exercise"* as the seed concept. This cycle continues until the virtual agent has determined that the learner's affect has already shifted to one that is more positive, or if the agent runs out of things to say.

Notice that in the given example, the options made by the user dictate the topic of discourse. Furthermore, the options do not necessarily relate directly to the reading material. This is intentional and is part of the storytelling process of providing opportunities for children to transfer the language they learn from stories to other personalized contexts. It is also a strategy to disrupt the user's negative affect to one that is more positive by presenting materials that a particular user may find interesting and appealing.

While Carla tries to appeal to topics that the user may find interesting, she must also ensure that the dialogue do not stray too far from the initial topic of discourse. Thus, the dialogue planner utilized by Carla performs recursive searches for semantically related concepts up to a level of three.

Another factor that Carla takes into consideration when choosing a topic of discourse is the concept's affective information. This is sourced from the polarity values stored in SenticNet [17]. Each commonsense concept in Carla's ontology is associated with a polarity value that describes how positive or negative is the affect being expressed by the concept. Carla should try to choose concepts with positive polarity values as a means of possibly shifting the learner's affect from negative to a more positive one.

Because detecting the learner's affect is outside the scope of the current research, the user has to explicitly inform Carla regarding his/her affect state anytime during the reading comprehension exercise or dialogue exchange by selecting either the "happy" or the "sad" icons. The dialogue then ends with an encouraging note to the learner to resume the learning activity, e.g., *"You seem to be feeling a lot better now. Maybe you can try working on the activity again? I'm sure you can do it now!"*.

4 Test Results

Table 5 shows the evaluation results from the preliminary testing of Ellie that was conducted among five (5) children diagnosed with mild autism. Children with mild autism are most capable of interacting with computer systems without requiring much assistance. They are able to live independently and possess a normal intelligence level, though they still struggle with decision-making tasks, common autism behavioral concerns such as overwhelming passion and interest towards and single topic, and social interaction and decorum.

Each participant was asked to go through the same story template five times to validate if he/she exhibited any learnings through the change in his/her choices. During this reading, the child should also interact with at least two objects or non-playing characters. Every option that has been selected is recorded. A shadow teacher is present to interpret the facial expressions or reactions of the child while using the system.

From the results in Table 5, the children seemed to have a preference for the teacher role of Ellie when they all gave a positive response in item #2. The facilitator role, on the other hand, only received affirmation from four of the children because the fifth child had difficulty understanding the instructions given by the peer, as seen in item #3. All children liked the stories that they read (item #4), although two of the participants struggled with comprehension issues. Specifically, the language used in the stories, English, is not the primary language of the children who participated in the test. These children had to be given an ample amount of time to read and be told the stories before they were able to grasp the underlying meaning.

Table 5. Evaluation results of Ellie.

Questions	Yes	Somewhat	No
1. Did you understand what Ellie said?	3	2	–
2. Was Ellie able to tell you what is right from wrong?	5	–	–
3. Did you understand the instructions?	4	–	1
4. Did you like the stories?	5	–	–
5. Were you able to understand the stories?	3	2	–
6. Did you learn something new from the stories?	5	–	–

The results, however, do not provide any validation as to the effect of the Ellie's roles to student learning. Furthermore, two of the participants had to seek assistance from the shadow teacher in order to understand their dialogue with Ellie. This necessitates the need for the system to be used under the supervision of a human teacher or guardian.

Carla, on the other hand, was evaluated by 13 students. Table 6 shows the results when Carla was evaluated based on its dialogue content only.

Table 6. Evaluation results for Carla's dialogue content.

Questions	Yes	Maybe	No
1. Did you understand the tutor?	9 (69.2%)	4 (30.8%)	–
2. Did the tutor provide choices that make sense?	6 (46.2%)	7 (53.8%)	–

From the results, 69.2% of the learners had no trouble understanding the words used by Carla, while the remaining 30.8% partially understood the tutor. This can be attributed to the manual process of populating the semantic ontology by extracting relevant assertions from ConceptNet, and the use of predefined templates to generate the dialogue content.

For the appropriateness of the list of candidate responses that Carla provides the user, 53.8% of the participants reported concerns that range from amusing choices to choices that do not make sense. Repetitive choices are also present, such as "*flower in the park*" and "*flower at a park*". Carla also sometimes produce redundant statements, such as "*you need an ice cream so you can eat an ice cream*".

Separate system testing showed that, given the current size of the knowledge base, Carla can engage the learner in a 20-turn dialogue (20 dialogue questions), though a number of the questions were already repeating.

5   Conclusion and Further Work

We presented the design of the semantic ontology of two virtual agents – Ellie who uses interactive storytelling to teach learners about social interactions; and Carla who engages its learners in a text-based dialogue during a learning activity. Both the ontologies adapted the semantic relations from ConceptNet while supplementing these with additional knowledge to ensure that the generated text contains concepts relevant to the story themes.

The combined use of commonsense ontology and template-based story generation provided a platform for Ellie to produce a variant of stories that adhere to the narrative structure of a Social Story™ while conforming to the language needs of the target audience. In the case of Carla, this approach allowed for the generation of lengthy dialogue exchange between the agent and the learner.

To address redundancy issues in Carla, further work should consider adding a concept attribute that will be used to determine the next assertion to be expressed in the dialogue. The attribute must be dynamic, such that its value changes as the dialogue progresses. This is to prevent the dialogue planner from selecting the same set of assertions for a given set of conditions. A possible attribute is the frequency in which an assertion is expressed relative to the number of dialogue turns that have already been exchanged between Carla and the learner.

Furthermore, Carla engages the learner in only one type of dialogue, specifically information-oriented dialogue turns that ask the learner about his/her knowledge or preferences for a given topic. Generating other types of dialogue, such as persuasion or negotiation to encourage the learner to go back to his/her learning activity, has not been considered in the current implementation. Further research on motivational factors in learning and how these can be used by the dialogue planner to generate other types of dialogue will be explored in the future.

While resources such as ConceptNet are readily available from the Web, the specific requirements of the learning environments presented in this paper necessitated the need to build separate semantic ontologies. This presents a problem since the task of populating the ontologies is mostly manual and time-consuming. Future works should explore related studies in the automatic or semi-automatic population of knowledge from corpus.

Ellie was validated among a very limited population of children with mild autism. Though the results showed the potential of Ellie as a virtual companion to help children comprehend the events in their daily lives and to prepare for social interactions, the tests did not provide conclusive evidence on the long-term effectiveness of the system as a learning environment.

Carla was validated among a slightly higher number of children compared to Ellie. Still, the test results did not provide any evidence relating to specific learning tasks and behavior, such as how the learner's bias towards the reading material may be affecting his/her attitude on the tutor's attempt at intervention, and the rationale for a conversational agent if the available reading materials are familiar to the target audience (and thus, they may have had positive affect throughout the reading activity).

Currently, both Ellie and Carla are visually portrayed as 2D graphical faces with fixed positive and negative facial expressions that are displayed at appropriate points based on the story plot. Future work can explore the use of animated virtual peers that can vary their expressions and possibly show an increasing capacity for affect that is visually authentic and appropriate to the ongoing discussions and interactions with the human user.

References

1. Cassell, J., Tartaro, A., Rankin, Y., Oza, V., Tse, C.: Virtual peers for literacy learning. Educ. Technol. Spec. Issue Pedagogical Agents **47**, 39–43 (2005)
2. Dautenhahn K., Davis, M., Ho, W.: Supporting narrative understanding of children with autism: a story interface with autonomous autobiographic agents. In: Proceedings of the IEEE International Conference on Rehabilitation Robotics, Kyoto International Conference Center, Japan, pp. 905–911. IEEE (2009)
3. Brophy, S., Biswas, G., Katzlberger, T., Bransford, J., Schwartz, D.: Teachable agents: combining insights from learning theory and computer science. In: Lajoie, S.P., Vivet, M. (eds.) Artificial Intelligence in Education, pp. 21–28. IOS Press, Amsterdam (1999)
4. D'Mello, S.K., Lehman, B., Graesser, A.C.: A motivationally supportive affect-sensitive AutoTutor. In: Calvo, R., D'Mello, S.K. (eds.) New Perspectives on Affect and Learning Technologies, vol. 3, pp. 113–126. Springer, New York (2011)
5. Ryokai, K., Vaucelle, C., Cassell, J.: Virtual peers as partners in storytelling and literacy learning. J. Comput. Assist. Learn. **19**(2), 195–208 (2003). Wiley Online Library
6. Baker, T.: Collaborative learning with affective artificial study companions in virtual learning environment. Ph.D. Dissertation, The University of Leeds (2003)
7. Xu, Y., Park, H., Baek, Y.: A new approach toward digital storytelling: an activity focused on writing self-efficacy in a virtual learning environment. Educ. Technol. Soc. **14**(4), 181–191 (2011)
8. Liu, H., Singh, P.: MakeBelieve: using commonsense knowledge to generate stories. In: Proceedings of the 18th National Conference on AI, pp. 957–958. National Conference on Artificial Intelligence, Edmonton, Alberta (2002)
9. Kim, K., Picard, R., Lieberman, H.: Common sense assistant for writing stories that teach social skills. In: Proceedings of CHI EA 2008 Extended Abstracts on Human Factors in Computing Systems, pp. 2805–2810. ACM, New York (2008)
10. Cua, J., Ong, E., Pease, A.: Using SUMO to represent storytelling knowledge. Philippine Comput. J. **5**(2), 37–43 (2010). Computing Society of the Philippines
11. Ong, E.: A commonsense knowledge base for generating children's stories. In: Proceedings of the 2010 AAAI Fall Symposium Series on Common Sense Knowledge, pp. 82–87, Virginia, USA. AAAI (2010)
12. Liu, H., Singh, P.: Commonsense reasoning in and over natural language. In: Negoita, M.Gh., Howlett, R.J., Jain, L.C. (eds.) KES 2004. LNCS, vol. 3215, pp. 293–306. Springer, Heidelberg (2004). doi:10.1007/978-3-540-30134-9_40

13. Singh, P., Lin, T., Mueller, E., Lim, G., Perkins, T., Zhu, W.L.: Open mind common sense: knowledge acquisition from the general public. In: Proceeding of the 2002 Confederated International Conferences DOA, CoopIS and ODBASE, pp. 1223–1237 (2002)
14. Gray, C.: The New Social Story Book, 10th edn. Future Horizons, Arlington, Texas (2010)
15. Groden, J., LeVasseur, P., Diller, A.: Picture This! Autism Spectrum Quarterly, pp. 18–21 (2007)
16. Speer, R., Havasi, C.: Representing general relational knowledge in ConceptNet 5. In: Proceedings of Eighth International Conference on Language Resources and Evaluation (LREC 2012), pp. 3679–3686. European Language Resource Association (2012)
17. Cambria, E., Speer, R., Havasi, C., Hussain, A.: SenticNet: a publicly available semantic resource for opinion mining. In: Proceedings of the 2010 AAAI Fall Symposium Series on Common Sense Knowledge, pp. 14–18. Association for the Advancement of Artificial Intelligence (2010)

Contents Organization Support
for Logical Presentation Flow

Tomoko Kojiri[1(✉)] and Yuta Watanabe[2]

[1] Faculty of Engineering Science, Kansai University, Suita, Japan
kojiri@kansai-u.ac.jp
[2] Graduate School of Science and Engineering, Kansai University,
3-3-35, Yamate-Cho, Suita, Osaka 5648680, Japan

Abstract. In appropriate presentations, each topic must be explained logically. To prepare appropriate presentation slides, all of the topics related to the research must be fully elucidated. The objective of this research is to help presenters derive enough topics that cogently explain their research theme. This paper proposes the logical model for the topics in the research presentation of the computer science field. Then, content organization support system that helps presenters organize topics based on the logical model is developed. Based on the experiment, our system was effective for creating a new contents and for organizing contents.

Keywords: Content map · Causal relation · Presentation support · Logical presentation

1 Introduction

We researchers often introduce our research results as presentations. In them, we often prepare presentation slides in which discussion topics are sequentially arranged. In such presentations, the order of topics is critical so that audiences correctly understand what the presenter is arguing. If the relations among topics are difficult to grasp, audiences might get confused. In this research, we define an appropriate presentation as one in which its topics are easy to understand, and an inappropriate presentation is difficult to understand.

Of course, such presentation tools as PowerPoint, Keynote, or Prezi provide functions that support the creation of presentation slides. However, as Kohlhase argued, these functions minimize the effort and time to create beautiful and stylish slides [1] and provide layouts or animation functions to create visually understandable slides. However, they do not support the preparation of contents for appropriate presentations. Several researches have encouraged presenters to improve their slide creation skills. One approach provides an environment in which comments about a presentation's good and bad aspects are simply gathered and exchanged during rehearsals [2, 3]. In this approach, whether a presenter can improve her presentation skills depends on who provides feedback, so presenters are not always able to improve their skills.

In appropriate presentations, each topic must be explained logically. That is, all topics must be connected by logical relations that are clearly represented in the slides.

© Springer International Publishing AG 2017
M. Numao et al. (Eds.): PRICAI 2016 Workshops, LNAI 10004, pp. 77–88, 2017.
DOI: 10.1007/978-3-319-60675-0_7

To prepare appropriate presentation slides, all of the topics related to the research must be fully elucidated. Their relations with other topics must be considered; logically-connected topics should be selected that adequately explain the presentation theme in the given presentation conditions.

This research focuses on two steps: deriving topics and considering relations between topics. The objective of this research is to help presenters derive enough topics that cogently explain their research theme. In research presentations, fundamental topics should be included with which the research is composed, such as goals, background, or methods. In addition, other topics are needed that explain the validity and reliability of each topic. Such additional topics may have relationships with fundamental topics. Our research develops a system that encourages presenters to derive both fundamental and additional topics and arrange them based on logical relations. For the purpose of encouraging users to induce new ideas, several idea processing systems provide related information with already derived ideas [4, 5]. Such systems do not consider necessary types of topics. Another approach helps presenters learn presentation slide composition based on the slides of experienced presenters [6, 7]. Types of topics and their relations may be different according to the research themes. Thus, this approach is not appropriate if the presentation themes are not identical to the experienced presenters' ones.

Our research does not directly support presentations. Instead, it encourages presenters to reflect on their own research and organize the topics that explain it. If topics are organized logically, creating appropriate presentations can be simplified by selecting required both fundamental and additional topics that have logical relations to the fundamental topics.

2   Logical Model for Research Presentations in Computer Science

A logical presentation explains its necessary topics with reasons. In research presentations, several explanations must be made, including the importance of the research's goal based on existing facts or opinions, the necessity of the proposed method by explaining how it achieves the goal, or its effectiveness by showing its evaluation results. To explain a presentation's topics, each one must be connected with reasons.

To support logically organized presentations for computer science researches, both fundamental and additional topics must be determined as well as their relations. Although Tanida et al. classified topics in the research presentations of computer science and proposed presentation semantics [4], their classification is too fine. Some of their classifications, such as logical approach and technological approach, are not necessary to create logical presentations. We eliminated some of the types and defined 11 types, which contain both fundamental and additional topics that can logically explain the fundamental topics.

First, we proposed a logical model of presentation contents in computer science (Fig. 1). In this model, links represent causal relations and nodes inside a node indicate inclusive relations. Research presentations propose methods to accomplish goals. Therefore, method and goal are necessary components. Since method is derived to accomplish a goal, a causal relation is attached from goal to method. Since methods sometimes consist of several sub-methods, methods have inclusive relations. To describe a goal's validity, its background is explained that elucidates the target problem that must be solved or the requirements that must be satisfied in the research. A goal sometimes becomes the background of another goal, so they have causal relations in both directions. The validity of goal and background or the effectiveness of the proposed method is evaluated by an experiment called an evaluation. An evaluation consists of several factors: what to evaluate (objective), how to evaluate (e-Method), by whom and where the experiment was conducted (environment), and so on. Based on the experimental results, sometimes new problems are discovered, which become the background of other researches. In addition to these components, we add reason components for causal relation links. A reason is a special component that explains why the presenter believes that two nodes share a causal relation. This may be an important component since it reflects a presenter's belief.

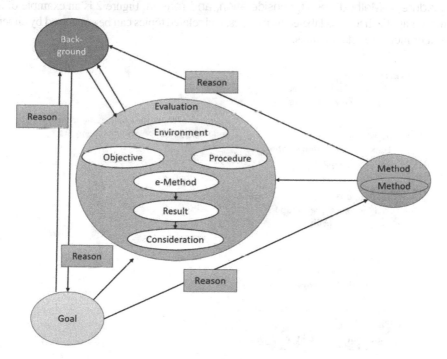

Fig. 1. Logical model for research presentations for computer science field

In our model, causal relations are not attached between topics. Background does not become a reason for a method, because no purpose was mentioned for deriving it. In addition, method is not a reason for establishing a goal, since a goal should solve the problems that are described in the background.

3   Content Map and Contents Organization Support Function

Since the topics that a presenter wants to discuss are implicit, the current topic structure, which a presenter has in his mind, is clarified and suggestions can be given for logical presentations if he externalizes his topics and their relations. By externalizing implicit knowledge, the presenter himself will notice the inappropriateness or incorrectness of his understanding of the topics.

In this research, we developed a system that (1) externalizes enough topics and (2) organizes them based on our logical model. In our system, we introduce a concept map [8], which is an externalization tool called a content map. Content maps consist of nodes and links. Nodes represent topics and links show the relations among topics. Based on a logical model, causal and inclusive relations are prepared as links. We set 11 types of nodes based on a logical model: background, goal, method, evaluation, environment, objective, procedure, e-Method, result, consideration, and reason. Figure 2 is an example of a content map. Each topic is labeled by its types, and related topics can be connected by either causal or inclusive relation links.

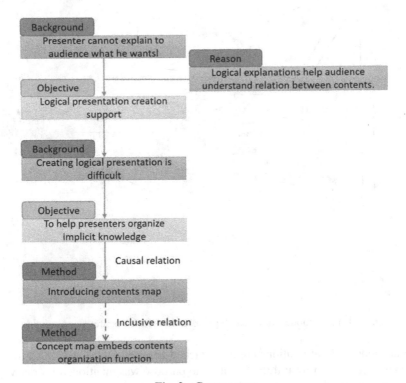

Fig. 2. Content map

A presenter sometimes is unable to create an appropriate contents map due to a (1) lack of topics and (2) incorrectly understanding relations. For such situations, a support function that organizes content map is useful. Several researches, which provide feedback about created concept maps [9], focus on learning activities to create correct content maps. For example, a teacher's concept map is prepared as a goal map and the differences between it and a student's concept map is given to students so they can clarify their understanding. In our research, goal maps cannot be prepared, since only the presenter knows what he wants to represent in his content map. Its validity can only be evaluated by the presenter himself.

To increase the awareness of presenters of inappropriate relations and a lack of topics, our content map embeds a contents organization function that highlights the available contents that can be connected by the causal/inclusive relations with selected nodes. If no nodes are highlighted, no link is generated. Our content map allows links between pairs of node types to which relations are defined in the logical model. Based on this contents organization support function, a presenter can understand the necessary relations between topics in a logical presentation by observing the highlighted node types. A presenter can also notice a lack of topics if he realizes that the expected topics are not highlighted.

Here, we explain the contents organization function in Fig. 3. Assume that Fig. 3(a) suggests an appropriate content map for a presenter to create. Figure 3(b) shows the current concept map that he created. When he wants to connect two goal nodes, our content map does not allow links between them since no links can be attached between goals in the logical mode. Based on this function, he might notice that he should create another node, such as background, to connect two goals.

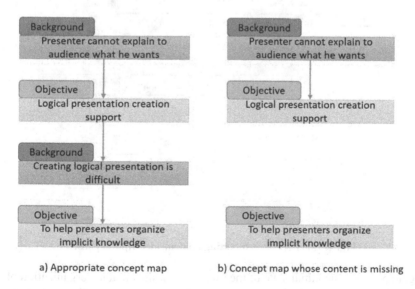

a) Appropriate concept map   b) Concept map whose content is missing

Fig. 3. Situation contents organization support function

4   Content Map Organization Support System

We have developed a system that embeds content maps and a contents organization support function. This system, which is implemented in C# programming language, consists of two phases to create content map: contents description and contents organization. In the contents description phase, a user writes topics and annotates them. In the contents organization phase, a user adds relations to the created contents and arranges all of the contents as a map. A user can rearrange these phases until all the topics are organized as a map.

Figure 4 shows the interface that supports the contents description phase. The contents slot represents individual topics, which consist of content sentences and annotation. A contents slot is added when the add button is pushed. Contents are created by adding a contents slot, filling in the contents sentences, and adding annotations. When the delete button next to the contents slot is pushed, the contents slot is deleted. When the next phase button is used, the interface for the contents organization phase appears and users can change to the next phase.

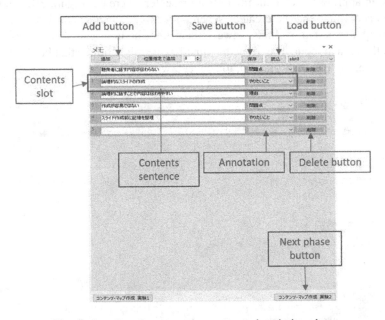

Fig. 4. Interface for supporting contents description phase

Figure 5 shows the initial state of the interface for the contents organization phase. Contents slots are transformed into contents nodes and arranged in the contents organizing area. Different colors are added to the contents nodes based on the attached annotations. For instance, backgrounds are depicted as red and methods as blue. To help a presenter recognize the meaning of these colors, explanations are shown on the top of the interface. In this interface, a presenter can drag and drop a contents node or remove a contents node by a right-button click with her computer's mouse.

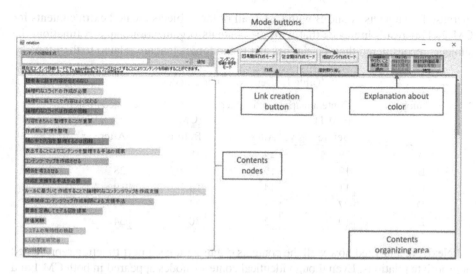

Fig. 5. Interface for supporting contents organization phase

A presenter changes modes to attach different relations. If a causal relation creation mode is selected, she can add causal relation links by selecting two contents nodes and pushing the link creation button. As in the contents organization support function, when selecting one contents node, contents nodes that have a causal relation with the selected one are underlined as available contents (Fig. 6). By selecting the second node from these underlined contents nodes, a link is attached that corresponds to the causal relation. However, if the contents node is selected without being underlined, no link is attached. In the same way, an inclusive relation link is attached in the inclusive relation creation mode. Also, a reason for the link is attached by selecting types of contents and target links.

5   Experiment

We experimentally evaluated our system's usability and the effectiveness of the contents organization support function. Four undergraduates and one graduate student (A to E) created content maps of their research topics in the following three steps. All five subjects had done research in the computer science field for over a year with little experience making presentations.

Step 1: Create content maps using a comparison system (CM 1). In it, subjects can create contents nodes, but they cannot add annotations. No contents organization support function is embedded. That is, subjects can connect any two contents nodes without any restrictions.

Step 2: Create their content map again using our content map organization support system (CM 2). In this phase, they can see the content map that they created in Step 1.

Step 3: Answer questionnaires.

First, we discuss the effect of the contents organization support function for deriving contents nodes. Table 1 shows the number of contents nodes that were derived before and after creating each content map. Although the number of contents nodes did not increase for subjects A and B with CM 1, all of the subjects created extra contents for CM 2. This result indicates that with the contents organization support function, the subjects reconsidered their researches and found topics that are related to their topics.

Table 1. Derived contents nodes

Subjects	Content map created in Step 1 (CM 1)		Content map created in Step 2 (CM 2)	
	Before	After	Before	After
A	29	29	28	37
B	19	19	27	28
C	11	14	14	19
D	27	31	34	39
E	19	22	30	34

Next, we analyzed how well the contents organization support function represented appropriate relations. Even though identical contents nodes appeared in both CM 1 and CM 2, their relations are not identical. The following are the types of relation changes between the contents nodes that appeared in both CM 1 and CM 2:

Type 1: Relation in CM 2 was identical as in CM 1.
Type 2: New contents node was added in CM2 between two connected contents nodes in CM 1.
Type 3: Type of relations was changed.
Type 4: Direction of type was reversed.
Type 5: Relation was removed.

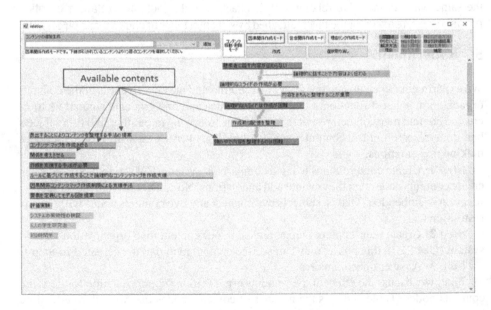

Fig. 6. Interface for indicating available contents

Table 2 shows the number of relations that correspond to these five types. For Types 2, 3, and 4, relations that were improved in CM 2 are shown in parentheses. Whether their relations improved was determined by the authors who were familiar with the research of these subjects. Pairs of contents nodes whose relations changed are Type 1, meaning that the subjects appropriately organized the contents nodes in CM 1. Only four relations of Type 2 were not improved: three relations of subject A and one relation of subject B. All other relations were appropriately changed while using the contents organization support function. Figure 7 shows an example of improved relations from CM 1 to CM 2. Since the content maps are written in Japanese, we attached English translations to each node. In CM 1, a causal relation link was attached from node a to node b. This relation was inappropriate, because both nodes were methods, and the goal that method b addressed was not mentioned. On the other hand, since background d, which was derived by method a and goal c, was added in CM 2, the reason for proposing method b clearly appeared. In this example, most relations were improved using contents organization function for Types 2, 3, and 4. This suggests that the contents organization support function is effective for logically arranging contents.

Table 2. Types of relation changes: numbers in parentheses correspond to numbers of appropriately changed relations

Subjects	Type 1	Type 2	Type 3	Type 4	Type 5
A	0	4 (1)	1 (1)	1 (1)	7
B	4	0 (0)	6 (6)	0 (0)	2
C	2	3 (3)	3 (3)	0 (0)	0
D	5	4 (3)	5 (5)	2 (2)	5
E	1	3 (3)	2 (2)	0 (0)	2

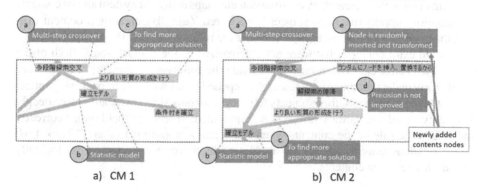

a) CM 1   b) CM 2

Fig. 7. Example of improved content map (Subject C)

Figure 8 shows the part of CM 2 that was changed by Type 5. Subject D wanted to connect consideration f and goal g. In this case, consideration f contains a problem, so he regards it to be a background node. However, he failed to connect them by a causal relation link in CM 2 because they are not directly connected in the logical model. In our current logical model, problems that are derived by evaluations should be represented background, not consideration. However, in some cases, problems are sometimes contained in considerations. To cope with this drawback, two annotations can be added to one content. If the consideration contains a problem, both consideration and background should be attached to the contents node.

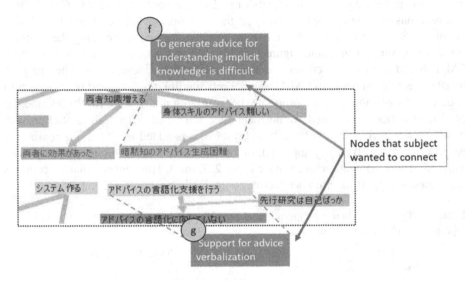

Fig. 8. Example of contents nodes for which subject failed to make a relation in CM 2 (Subject D)

Table 3 indicates the questionnaire results and shows the numbers of subjects who selected individual items. Three subjects believed that they increased their understanding of their own research by creating content maps using our system and its contents organization support function. Subject E answered Yes: "By creating a content map with the system, I was encouraged to consider the relations among the existing contents and to find extra contents that were not externalized in the content map." Both of the subjects who answered No in question 1 made the following comment: "I'm not sure whether my understanding of my research deepened, but I was able to positively reflect on it." On the other hand, four subjects felt they understood their research more deeply when they added annotations. Subjects A commented: "I realized that I didn't correctly understand the role of the contents when I attached annotations to them." Thus, both adding relations among contents and attaching annotations were effective for logically considering research contents.

Table 3. Questionnaire results

Questions	Yes	No
1. Did you understand your research more deeply by creating content map with our system?	3	2
2. Did you understand your research more deeply by attaching annotations?	4	1

6 Conclusion

This research supported presenters to create more logical presentations by organizing their topics. This paper proposed a logical model for topics in the research presentations of the computer science field and developed a concept map with a contents organization support function. Based on our experiment, our system effectively created new contents and organized them. However, we haven't yet evaluated the validity of our logical model. We must evaluate whether it appropriately represents the logical relations among topics in the computer science field.

Our system currently limits the available pairs to create links to encourage presenters to identify appropriate pairs for causal and inclusive relations. During evaluations, some presenters failed to create links because they could not derive appropriate types of nodes. To support such presenters, providing advice for generating specific types of contents might be effective. Therefore, future work will develop a function that determines the type of missing contents and encourage presenters to positively derive them.

Acknowledgement. The work was supported in part by JSPS KAKENHI Grant-in-Aid for Scientific Research (B) (No. 16H03089) and JSPS KAKENHI Grant-in-Aid for challenging Exploratory Research (16K12563).

References

1. Kohlhase, A.: Semantic PowerPoint: content and semantic technology for educational added-value services in MS PowerPoint. Proc. of World Conference on Educational Media and Technology, pp. 3576–3583 (2007)
2. Okamoto, R., Tanikawa, A., Kashihara, A.: Presentation authoring support considering relationship between slide contents and oral expressions for peer review in presentation rehearsal. In: Proceedings of eLearn2015, pp. 996–1001 (2015)
3. Kojiri, T., Nasu, H., Maeda, K., Hayashi, Y., Watanabe, T.: Collaborative learning environment for discussing topic explanation skill based on presentation slides. In: Proceedings of 12th European Conference on e-Learning (ECEL 2013), vol. 1, pp. 199–208 (2013)
4. Kanakubo, M., Hagiwara, M.: "Creativity support system combining morphological analysis method and modified input-output method. In: Joint International Conference on Soft Computing and Intelligent Systems and 3rd International Symposium on Advanced Intelligent Systems (2002)
5. Nishimoto, K., Abe, S., Miyasato, T., Kishino, F.: A system supporting the human divergent thinking process by provision of relevant and heterogeneous pieces of information based on an outsider model. In: Proceedings of International Conference on Industrial and Engineering Applications of Artificial Intelligence and Expert Systems, pp. 575–584 (1995)
6. Tanida, A., Hasegawa, S., Kashihara, A.: Web 2.0 services for presentation planning and presentation reflection. In: Proceedings of International Conference on Computers in Education, pp. 565–572 (2008)
7. Shibata, Y., Kashihara, A., Hasegawa, S.: Scaffolding with schema for creating presentation documents and its evaluation. In: Proceedings of World Conference on E-Learning in Corporate, Government, Healthcare, and Higher Education, pp. 2059–2066 (2012)

8. John, R.: McClure: "Concept map assessment of classroom learning: reliability, validity, and logistical practicality". J. Res. Sci. Teach. **36**, 475–492 (1999)
9. Hirashima, T., Yamasaki, K., Fukuda, H., Funaoi, H.: Kit-build concept map for automatic diagnosis. In: Biswas, G., Bull, S., Kay, J., Mitrovic, A. (eds.) AIED 2011. LNCS, vol. 6738, pp. 466–468. Springer, Heidelberg (2011). doi:10.1007/978-3-642-21869-9_71

A Framework to Generate Carrier Path Using Semantic Similarity of Competencies in Job Position

Wasan Na Chai[1(✉)], Taneth Ruangrajitpakorn[1,2], Marut Buranarach[1],
and Thepchai Supnithi[1]

[1] Language and Semantic Technology Laboratory, National Electronics
and Computer Technology Center, Pathumthanee, Thailand
{wasan.na_chai, taneth.rua, marut.bur,
thepchai.sup}@nectec.or.th
[2] Department of Computer Science, Faculty of Science and Technology,
Thammasat University, Pathumthanee, Thailand

Abstract. A career path is necessary for students and workers to keep themselves in track for their career goal. However, a career path following a job standard in general is very rare. This paper presents a method to find a semantic similarity within competencies of Job positions for realising a path to relate career. By a development of Thai WordNet containing terms used in competency description, a distance of classes of WordNet structure is used to determine a semantic similarity of competencies. Paths to relate job positions are assumed for the job positions sharing similar competencies, and the more they share, the more transferrable job is viable. From the usage scenario, the proposed framework proved that semantic of words is more useful than using character based similarity in competency comparison.

Keywords: Career path · Semantic similarity · WordNet · Competency

1 Introduction

A career path is important for students and workers for planning their career towards their goal in life. In a career path, job positions are linked to other in two types, i.e. promotion and transferring job. A promotional path is a possible direction of going higher in the same job position while a transferring path is to change to another job branch. Moreover, changing a job is commonly necessary since people can be late-bloomer to find their own potential or become boring of the job. Changing to a job requiring competencies that a person already has will give him/her a good advantage in his/her new career. It is common in a company or an organization to provide a promotional path with certain criteria for its employees, but the job-transferring path is scarcely available.

There was a work mentioned on automatically generating a career path from a qualification data [1]. The work exploited certification names related to a job and their assigned competencies from Thailand Professional Qualification Institute (TPQI) [2] for linking paths to form a career path. By comparing certification names and competency

© Springer International Publishing AG 2017
M. Numao et al. (Eds.): PRICAI 2016 Workshops, LNAI 10004, pp. 89–97, 2017.
DOI: 10.1007/978-3-319-60675-0_8

names, they mainly exploited a string similarity to find commonness and assigned linking to similar concepts. They can demonstrate a career path based on a string similarity. From the result, there were some missing paths from the obtained career path since string similarity apparently overlooked the terms that are synonym or semantically related.

This paper aims to present an upgraded version of a framework to generate a career path by focusing on semantic of terms in a competency to increase coverage over the existing work. We expect to capture a semantic similarity in competencies in a different surface form of terms and create a relation between competencies. With similar competencies, job positions in a different career will be assigned with a path to inform as a transferrable job.

2 Background

2.1 Professional Qualification Standards

Professional or vocational standards or competency standards are the standards of performance individuals must achieve when carrying out functions in the workplace, together with specifications of the required knowledge and understanding [3]. They describe what an individual needs to do, know and understand in order to carry out a particular job role or function in his or her occupation. They are normally defined by a representative sample of employers and other key stakeholders and approved by the national qualification standard organization. Many countries have established occupational competence standards, to support skills development, employability and vocational education developments. This is to help enabling skills transfer and recognition, supporting skills credibility, and global economic competitiveness.

Professional qualification is a certification earned by a person to assure qualification to perform a job or task. They typically involve competence-based assessment. Competencies include all the related knowledge, skills, abilities, and attributes that form a person's job. Identifying employee competencies can contribute to improved organizational performance. Competency assessment criteria are normally based on national professional qualification standards. Qualification levels defined by the standards should be comparable between different industries. For example, a level 2 competency qualification in engineering will be a comparable achievement to a level 2 competency qualification in construction. Figure 1 shows relationship between competency, qualification of vocational competence and assessment criteria.

Fig. 1. Relationship between competency, qualification of vocational competence and assesment criteria

2.2 Related Work

Previous work [1] presents a methodology for automatic career path generation based on certifications and their competencies provided by TPQI. The work attempted to find a path among task/role by using a similarity of characters in the certification names and competencies. The similarity score of this work is calculated by formula (1).

$$NMLCS_1(X_{i-1}, Y_{j-1}) = \frac{length\left(NMLCS_1(X_{i-1}, Y_{j-1})\right)^2}{length(X_{i-1}) \times length(Y_{j-1})} \tag{1}$$

By comparing common characters, the competencies indicated as similar were mapped together as a path to another job position. A percentage of non-union number of competencies is calculated to show a number of competences required to transfer to another career. By the experiment, this work shows that the method work fine in career path generation, but some path result are not accurate since the some paths were over-generated from long length similar characters although the competencies are not semantically related.

3 Methodology

This paper aims to find semantic similarity among units of competency (UoCs) for assigning a path to jobs in different career. Terms representing content (content word) are gathered and categorised into a hierarchical structure based on WordNet [4] formation as called Competency Net (CompNet). According to WordNet, this CompNet consists of a set of synonym terms called SynSet, and these SynSets are hierarchically structured in hypernym-hyponym relation. With CompNet, a semantic similarity of terms in UoC can be calculated. An overview of the framework is illustrated in Fig. 2.

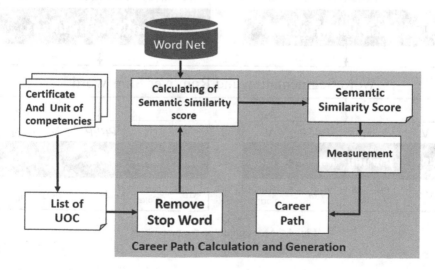

Fig. 2. An overview of the proposed framework

3.1   Pre-processing

Data from TPQI are structured as a profession for a root. For each profession, certifications are assigned to represent job position along with its level. Every certifications compose of units of competency (UoCs) which are all required to obtain a qualification for a certification. A schema of the data structure of TPQI is sketched in Fig. 3.

Fig. 3. A schema of TPQI data

An input of the framework is a list of UoCs belonging to a certain certification. UoCs are given as a Thai text describing skill and knowledge required for a job position. Since a key to inform technical skill and knowledge relies on with noun and verb, they are extracted by exploiting a word segmentation and POS tagger tool [5, 6]. Hence, only nouns and verbs in UoC will later be processed in the further steps. An example of UoCs and its output from the preprocess is shown in Fig. 4.

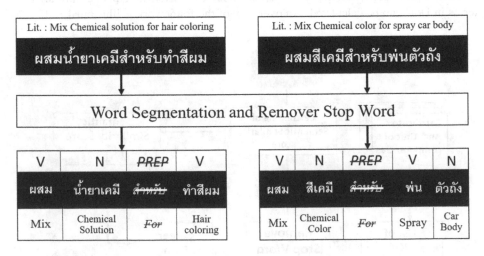

Fig. 4. An example of the preprocess

3.2   CompNet Development

CompNet is a lexical resource developed according to a design of Thai WordNet [7, 8]. The CompNet is designed as a representation of semantic existing specifically in UoCs. However, some feathers provided in WordNet are not used in this work; hence, they are ignored in CompNet development such as definition and form derivation. CompNet consists of Thai terms existing in UoCs, and the terms are categorised into a synonym set (SynSet) related to one another in terms of hypernym-hyponym relation. It was carefully crafted following a hierarchy of SynSets in WordNet and extended with the specific terms mentioned in UoCs. Some lexical entries of CompNet (with their literal translation for more understanding) are illustrated in Fig. 5.

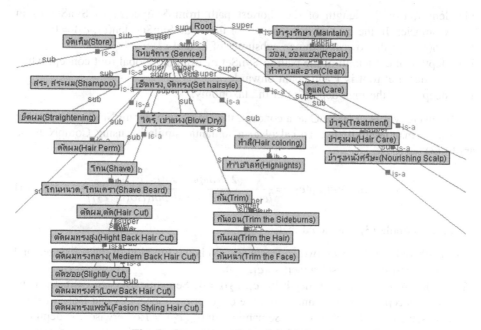

Fig. 5. Some parts of the defined CompNet

3.3   Semantic Similarity Calculation

From preprocess, nouns and verbs in UoCs are left over for calculating to semantic similarity score. In this work, we aim to compare an UoC with other UoCs of a certification of another profession to signify commonness in skill and knowledge to hint a transferrable path. Therefore, each word in UoC in a set will be compared to all other words of another UoCs from another certification. We exploit formula (2) to calculate a similarity score of a pair of UoCs.

$$Sim(UoC_1, UoC_2) = \frac{\sum_{i=1}^{n} Semantic\_Sim(\max(Wi_{UoC_1}, Wj_{UoC_2}))}{count(W_{total})} \qquad (2)$$

Where:

UoC is a competency to be compared for similarity

W is a word in a competency

n is a number of word and W_{total} is the total amount

The CompNet is used to measure a distance between terms. We apply a semantic similarity measure introduced by Wu Z. and Palmer M. [9, 10]. This semantic similarity measure focuses on the position of concepts in the taxonomy relatively to the position of the most specific common concept (their shared mother node). With the idea, the similarity between two concepts relies on the function of length and depth of the path.

Before further explanation, let's define notations as follows:

(1) len(c_1, c_2): the length of the shortest path from SynSet c_1 to SynSet c_2 in CompNet. In the case of the concepts of the same SynSet, len(c_1, c_2) = 0.
(2) lso(c_1, c_2): the lowest common subsumer of c_1 and c_2.
(3) depth(c_1): the length of the path to SynSet c_1 from the global root concept while the depth at root is 1 and count onwards.
(4) deep_max: the max depth (c_i) of the taxonomy.

By using words in an UoC as a concept, the structure of CompNet can be used to measure semantic similarity. To calculate a semantic similarity using CompNet, we apply (3).

$$Semantic\_Sim(c_1, c_2) = \frac{2 * depth(lso(c_1, c_2))}{len(c_1, c_2) + 2 * depth(lso(c_1, c_2))} \tag{3}$$

From formula (3) it is noted that,

(1) The similarity between two concepts (c_1, c_2) is the function of their distance and the lowest common subsumer(lso(c_1, c_2)).
(2) If the lso(c_1, c_2) is root, depth(lso(c_1, c_2)) = 1, Semantic_Sim(c_1, c_2) > 0; if the two concepts have the same sense, the concept c_1, concept c_2 and lso(c_1, c_2) are the same node. len(c_1, c_2) = 0. Semantic_Sim(c_1, c_2) = 1;otherwise 0 < depth(lso(c_1, c_2)) < deep_max, 0 < len(c_1, c_2) < 2*deep_max, 0 < Semantic_Sim(c_1, c_2) < 1.

In the case that a word amount from the comparing UoCs is different, Semantic_Sim of c_1 to c_2 and c_2 to c_1 will be combined and divided by 2 for normalising the score. A word pair with the highest Semantic_Sim score will be selected as aligned words from UoC_1 and UoC_2.

4   Usage Scenarios

This section mentions how the proposed method works against the string similarity score from the previous work [1]. The string similarity in comparison applies a calculation result from formula (1). There are two case scenarios.

The first scenario is the comparison of two UoCs shown in 2*3 matrix in Fig. 6. For traceability of an explanation, a related part of CompNet is also provided. This scenario shows that these UoCs gain a semantic similarity score for 0.88 while they obtain a string similarity score for 0.45. This case shows that these two UoCs share several words with a same semantic. As for a calculation of the proposed method, we first calculated Semantic_Sim score of each word from UoC_1 and UoC_2. For example, the first word of UoC_1 and the first word of UoC_2 gain Semantic_Sim as 1.0 since it found that they are the words in the same SynSet, and they are selected since they gain the highest score among all pairs. Another example is the third word of UoC_1 and words from UoC_2. The word is not in a SynSet with other words so it is calculated using (3). Since a length of the word to the comparing words is 5 (counting from nodes), and $depth(lso(c_1, c_2))$ is 1 since the lowest common subsumer of them is the root, the Semantic_Sim score is calculated as 0.29. By applying (3), the result of score is 0.76 (from (1.0 + 1.0 + 0.29)/3) for UoC_1 to UoC_2. However, the number of words in UoC_1 and UoC_2 does not equal; hence, a similarity score of UoC_2 to UoC_1 are required to be calculated for bidirectional normalisation, and we obtain 1.0 as the score. Hence, the similarity score of UoC_1 and UoC_2 is 0.88 from normalising of 0.76 and 1.0. The scenario shows that this method can relate a semantic of words in different UoCs effectively while the string similarity calculation from the previous work ignores this pair of UoCs.

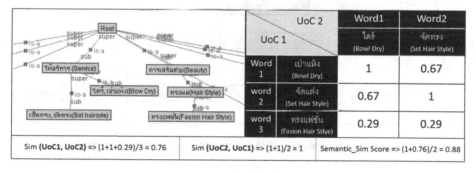

Fig. 6. Word matrix of first scenario and their details

The second scenario is the comparison of long UoCs. In this case, the words, which do not belong to POS noun and verb, are removed into 3*4 word matrix as shown in Figs. 7 and 8. The interesting words in this case are 'สีเคมี' (chemical colour) and 'น้ำยาเคมี' (chemical solution). These terms are related in CompNet in the same tree while *chemical solution* is a superclass of *chemical colour*. In this case, $len(c_1, c_2)$ of these words is 1, and $depth(lso(c_1, c_2))$ is 3 since the lowest common subsumer is the class *chemical solution* itself at 3 levels from root. By the combination of the first two words are very high, but the difference of the rest words reduce the total score to 0.66. By the meaning, these two competencies do not supposed to be apparently related since the UoC_1 is about hair dying, and UoC_2 mentions about car painting. With the help of

UoC 2 / UoC 1		Word1 ผสม (Mix)	Word2 สีเคมี (Chemical Color)	Word3 พ่น (Spray)	Word4 ตัวถัง (Body)
Word1	ผสม (Mix)	1	0.25	0.29	0.29
word2	น้ำยาเคมี (Chemical Solution)	0.29	0.85	0.25	0.25
word3	ทำสีผม (Hair Coloring)	0.25	0.25	0.29	0.29
Sim (UoC1, UoC2) => (1+0.85+0.29)/3= 0.76					
Sim (UoC2, UoC1) => (1+0.85+0.29+0.29)/4= 0.61					
Semantic_Sim Score => (0.76+0.61)/2 = 0.67					

Fig. 7. Word matrix of second scenario

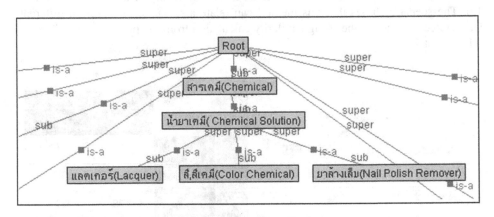

Fig. 8. Details of words and their CompNet for the second scenario

the rest words, these two UoCs are not recognised as related, but it could be better if we can recognise the difference in this specific technical terms automatically.

5 Conclusion and Future Work

In this paper, we propose a method to generate a career path using a semantic similarity of Thai words described in a competency for job position. To improve a capability of the previous work, semantic of the words is focused as the core for creating a path between job positions. CompNet (a hierarchy of words in competency following the concept of WordNet) was developed as a reference for semantic distance of the words. By calculating a score of words from different competencies, a total score of semantic similarity between UoCs is obtained. Paths to relate job positions are assumed for the job positions sharing a semantically similar competency. From the usage scenario, we found that the proposed method greatly improves the ability to generate a career path by exploiting semantic against the previous method using only string similarity. More paths that were never recognised by the prior method are formed by the semantic.

In the future, we plan to focus on finding the terms mentioning specifically technical skill, knowledge and attitude in a competency since these terms uniquely represent specialty in their relative field and should not be related to other competencies. Morcover, we will use the result of this method for visualisation to support Thai students and workers in achieving career goal.

References

1. Ruangrajitpakorn, T., Na Chai, W., Buranarach, M., Supnithi, T., Kongkachandra, R.: An automatic Thai career path generation using similarity of roles and their competencies. In: 2015 International Symposium on Multimedia and Communication Technology. Phranakhon Si Ayutthaya, Thailand (2015)
2. Thailand Professional Qualification Institute Homepage. http://www.tpqi.go.th/en/
3. National Occupational Standards (NOS) Database. http://nos.ukces.org.uk/
4. Miller, G.A., Beckwith, R., Fellbaum, C.D., Gross, D., Miller, K.: WordNet: an online lexical database. Int. J. Lexicograph. 3(4), 235–244 (1990)
5. SegIt: Thai Word Segmentation Tool. http://thaimt.org/lstnlp/wordseg.php
6. PosIt: Online Thai POS Tagger. http://thaimt.org/lstnlp/pos.php
7. Thoongsup, S., Robkop, K., Mokarat, C., Sinthurahat, T., Charoenporn, T., Sornlertlamvanich, V., Isahara, H.: Thai WordNet construction. In: the 7th Workshop on Asian Language Resources, ACL-IJCNLP, pp. 139–144. Singapore (2009)
8. Leenoi, D., Supnithi, T., Aroonmanakun, W.: Building a gold standard for Thai WordNet. In: International Conference on Asian Language Processing 2008, Chiang Mai, Thailand (2008)
9. Wu, Z., Palmer, M.: Verb semantics and lexical selection. In: Proceedings of 32nd Annual Meeting of the Association for Computational Linguistics. June 27–30, Las Cruces, New Mexico (1994)
10. Meng, L., Huang, R., Gu, J.: A review of semantic similarity measures in WordNet. Int. J. Hybrid Inf. Technol. 6(1), 1 (2013)

AI4T 2016: Workshop on Artificial Intelligence for Tourism

Inferring Tourist Behavior and Purposes of a Twitter User

Yuya Nozawa[1(✉)], Masaki Endo[1,2], Yo Ehara[3], Masaharu Hirota[4],
Syohei Yokoyama[5], and Hiroshi Ishikawa[1]

[1] Graduate School of System Design,
Tokyo Metropolitan University, Tokyo, Japan
{nozawa-yuya, endo-masaki}@ed.tmu.ac.jp,
ishikawa-hiroshi@tmu.ac.jp
[2] Division of Core Manufacturing, Polytechnic University, Tokyo, Japan
[3] National Institute of Advanced Industrial Science and Technology,
Tokyo, Japan
y-ehara@aist.go.jp
[4] Department of Information Engineering, National Institute of Technology,
Oita College, Oita, Japan
m-hirota@oita-ct.ac.jp
[5] Faculty of Informatics, Shizuoka University, Hamamatsu, Japan
yokoyama@inf.shizuoka.ac.jp

Abstract. The importance of tourism information such as tourism purposes and tourist behavior continues to increase. However, obtaining precise tourist information such as the tourist destination and tourism period is difficult, as is applying that information to actual tourism marketing. We propose a method to classify Twitter user into tourist behavior and tourism purposes, extracting related information from Twitter posts. Our experiments demonstrated a 0.65 F-score for multi-class classification, showing accuracy for inferring tourist behavior and tourism purposes.

Keywords: Attribute estimation · Travel information

1 Introduction

Recently, tourism occupies an important position in many countries as a key industry. Consumption activities related to tourism positively affect industries such as transportation, lodging, and manufacturing. Therefore, increasing the number of tourists is an important issue for governments and companies.

Providing tourists with sufficient information is demanded to increase tourism. For example, to obtain information related to tourist attributes and destinations, Japan's government has conducted surveys of tourism markets[1]. In this survey, inbound tourists state where they plan to sightsee, by questionnaire. The most important benefit of this approach is that the result has high reliability because the investigation method is face-to-face in almost all cases. Another advantage is that it is easy to obtain

---

[1] http://www.mlit.go.jp/kankocho/siryou/toukei/syouhityousa.html.

© Springer International Publishing AG 2017
M. Numao et al. (Eds.): PRICAI 2016 Workshops, LNAI 10004, pp. 101–112, 2017.
DOI: 10.1007/978-3-319-60675-0_9

adequate information because the questioner can set the contents of questions as intended. However, this questionnaire method presents some shortcomings. First, the amount of data is limited by time and monetary costs for researchers. Second, it is difficult to set questions to obtain sufficient information in line with what researchers want to analyze. Third, even if a questioner sets complete question items to obtain the necessary information, the analyzer cannot use it flexibly after obtaining the results because of the fine granularity of the resulting information. Therefore, it is difficult to use questionnaire results obtained using the investigation method.

Therefore, some researchers specifically examine social media sites to extract tourist information instead of survey by questionnaires. Tourists have been posting reports of their impressions and opinions related to tourist attractions to social media sites. In Twitter, users post the contents on the fly and note impressions that one might experience at a tourist destination. Extracting information from Twitter have some benefits that are not provided by questionnaires administered to extract tourism information. First, to the method obviates the dispatch of research workers for the survey. Second, the analyzer can obtain huge amounts of data at low cost. Finally, no need exists to determine in advance what to analyze. One can determine issues of granularity, such as time and space of the data examined, to meet the analytical goals. However, it is difficult to obtain information such as population and distribution of sampling from Twitter. Therefore, Twitter is not suitable for research such as statistics, but suitable for research such as impression, behaver and senses.

Table 1. Examples of Behaviors of "sightseeing" and "business"

sightseeing	tour, tourism, spa, nature viewing, museums, concerts, movies, theater appreciation, participation events, botanical gardens, theme parks.
business	visit such as the head office, branches and factories, suppliers visit, participation in training and seminars, professional sports activities.

Therefore, we propose a method to extract information related to tourist behavior and tourism purposes from Twitter. To infer tourist behavior, we classify each tweet posted during a tourism period to classes of tourist behaviors. Then, to infer the tourism purpose, we classify a group of tweets posted during a tourism period to classes of tourism purposes. We define a user who posted tweets another area from the biosphere during a short period of time. Then, we regard that tourism purpose as representing the reason why a user does tourist activities. Tourism purpose classes are "sightseeing", "business" and "other purpose". Additionally, in this research, tourism purposes of "business" and "sightseeing" are inferred according to Table 1 created by the tourism Purpose of the Travel and Tourism Consumption Trends Survey of the Tourism Agency[2] as a reference. In addition, tourist behavior represents what a user was doing at the time of posting a tweet. For this research, we set five classes of tourist behavior as "sightseeing", "business", "eat", "buy" and "other behavior". In our approach, we infer the tourism purpose and tourist behavior in separate steps. For example, a user

---

[2] http://www.mlit.go.jp/kankocho/siryou/toukei/shouhidoukou.html.

who goes on a business trip might post some tweets about liking eating and sightseeing. The tourism purpose should be classified as "business". However, each activity should not be classified as "business", and those are classified as "sightseeing" and "eat". In other words, the tourism purpose and tourist behavior show a gap on tourism information that a researcher can analyze. Therefore, when obtaining detailed information of the user related to tourism, one must elucidate both classes separately.

2 Related Research

Some researchers proposed the inference of user attributes. The main targets of estimation are gender [3, 4], age [4, 5], political-orientation [4, 6], residence [7], and occupation [8]. Classification of tourist behavior and tourism purpose to address in this research are also features of research related to the estimation of user attributes.

Research to extract tourist information from the information related to the Web has been conducted actively. Li et al. [9] proposed a method to divide travel purpose words into seven types, such as "scenic spots", "shopping", and "food". Zamal et al. [10] inferred "age", "gender", and "political preference" by combining estimation results and classifiers using the data and lexical identity of Twitter friend relationships. Ishino et al. [11] extracted traveler's transportation information automatically from travel blog entries written in Japanese using machine-learning techniques. Methods used in the estimation of user attributes are diverse. When classifying the attributes using supervised learning among them, Support Vector Machine (SVM) [12] is used particularly often. Benevenuto et al. [13] defined "spammers" who continue to send spam on Twitter. They are classified using SVM as "spammers" and "non-spammers". Pennacchiotti et al. [14] identify "political affiliation" and "particular ethnicity" and "Starbucks fans" from Twitter.

For this research, we propose a method to classify each tourist behavior as "sightseeing", "business", "eat", or "buy". The tourism purposes of "business" and "sightseeing" use SVM to target Twitter. A point of novelty of this research is that tourists of "business" purposes also do "sightseeing". To classify tourism only regarding purposes of "business" and "sightseeing", we classify the behavior what tourists are doing in tourism. In addition, there is a novelty even to the point of using multi-label SVM. We consider that the most tweets belong to more than one class.

3 Proposed Method

In this section, we describe a method to infer Twitter user tourist behavior and tourism purposes using tweets. First, we extract the tourism period of a Twitter user from the tweets. Second, to infer tourist behavior during the tourism period, we classify each tweet to their tourism classes based on the tweet text. Finally, to infer the tourism purpose, we propose two methods: one based on text classification using tweets, and one using aggregation of the inferred results of tourist behavior.

As described in this paper, we define five classes of tourist behavior with four behavior classes "sightseeing", "business", "eat", and "buy", and "other behavior". Additionally, we define two purpose classes as "sightseeing" and "business", and "other purpose". We define classes that represent other purposes for two reasons. First, sometimes, a user posts tweets without doing any tourism during travel period. Second, even after eliminating such noisy tweets, inferring appropriate classes is difficult because user behavior and purposes are widely varied.

3.1 Preprocessing

To obtain the tweets, we used the Twitter Streaming API[3]. At that time, we eliminated tweets that had been posted from countries other than Japan. Next, we apply preprocessing to the obtained tweets. We delete tweets including auto-generated texts from other social media sites. Additionally, we delete replies, retweets, URLs, and pictograms from the body of the acquired tweet.

3.2 Extraction of Tourism-Destination-Related Tweets

This section describes the procedure for extracting user tweets posted from the vicinity of a specific tourism destination. As described in this paper, we regard Tokyo as specific tourism destination. Therefore, we extract tweets of a Twitter user who stays outside Tokyo regularly, and who stays in Tokyo for a short time.

1. First, we sort the user tweets in chronological order. Additionally, we set a latitude–longitude bounding box of tourism destinations, and obtain all tweets that the user posted related to the tourism destination.
2. Second, we ascertain whether the user is a tourist or not. We assume that when the number of tweets posted at the tourism destination is sufficiently smaller than all tweets, then the tweets were posted during the travel period. Therefore, we calculate the percentage of the tweets posted at the tourism destination among all tweets. The user is defined as a tourist if the result is below the threshold. We tried each threshold as 0.1, 0.2, …, 0.5 and confirmed that the best threshold is 0.3 by visual inspection. Here, tweets posted at the tourism destination by the tourist are defined as a "tourism tweet".

In almost all cases, the tourism tweets are continuous in all tweets. Therefore, we combined the texts of consecutive tweets posted during a tourist period to produce a single document. We designate the document as the "tourism period document". Tourism tweets are used to infer the tourist behavior. In addition, the tourism period document is used to infer the tourism purpose using SVM.

---

[3] https://dev.twitter.com/overview/documentation.

3.3 Vectorization

This section describes a means of vectorizing texts of the tourism tweets and tourism period documents. First, we apply morphological analysis to the text of tourism tweets. As described in this paper, we use nouns, verbs, and adjectives from extracted morphemes. For this research, we use MeCab[4] as a morphological analyzer, and mecab-ipadic-NEologd[5] to morphological analyzer. Next, we apply tf-idf to the words. Then, we apply Latent Semantic Analysis (LSA) [15], to reduce the tf-idf dimensionality. The process of vectorizing tourism period documents is the same.

3.4 Filtering of the "Other Behaviors" and "Other Purposes"

In this section, we describe the method used to filter "other behavior" and "other purpose". Both other classes represent the noisy class. Tourism tweets that are not classified as "sightseeing", "business", "eat", and "buy" is defined as "other behavior". In addition, tourism period documents that are not classified as "sightseeing" or "business" are defined as "other purpose". In the following steps, we classify tourism tweets and tourism period documents into four and two classes. Therefore, to improve performance of the steps, we filter those other classes preliminarily.

We use Support Vector Machine (SVM) to filter others. We classify tourism tweets and tourism period documents into "non-other behavior" and "other behavior" by SVM using a manually generated training dataset. Similarly, we filter tourism period documents. In the following steps, we apply our method to tourism tweets and tourism period documents without others classified in this step.

3.5 Inference of Tourist Behavior

In this section, we describe a method to classify a tourism tweet into the four classes of "sightseeing", "business", "eat", and "buy". We use the Multi-label SVM to classification using a feature vector created in Sect. 3.3. We use multi-label SVM because we infer that a tweet often includes several behaviors. For example, the tweet "I finished meeting at Tokyo, and will go to lunch" includes classes "business" and "eat". For that reason, multi-label classification is a more suitable method for inferring tourist behavior than multi-class classification. Here, multi-label SVM, which we use, is implemented in a combination of two-class SVM.

3.6 Inference of Tourism Purpose

This section presents a method to infer the tourism purpose of a user in tourism period. Classes of tourism purpose are three: "sightseeing", "business" and "other purpose". We propose approaches of two types: using results of tourist behavior calculated in

---

[4] http://mecab.googlecode.com/svn/trunk/mecab/doc/index.html.

[5] https://github.com/neologd/mecab-ipadic-neologd.

Sect. 3.4; and using multi-class SVM based on tourism period documents. We present the following procedure for using tourist behavior results.

1. Documents are classified as "business" if at least one tourism tweet in a tourism period document was classified as "business". This is true because we assume that a tourist entering the country for business might do something corresponding to the class of "sightseeing" after business matters are completed. However, a sightseeing tourist does not correspond to "business".
2. For a tourism period document that does not contain the class of "business" in a tourism tweet, a document that includes "sightseeing" is classified as "sightseeing".
3. The other case, a tourism period document is classified into "other purpose".

Next, we describe another method using tourism period documents and multi-class SVM. We apply multi-class SVM to a tourism period document. Then, we classify the document as "sightseeing", "business", or "other".

4   Evaluation

4.1   Experimental Conditions

We use tweets posted in Japan posted during March 11, 2015 through October 28, 2015. By the processes described in Sect. 3.1 and Sect. 3.2, we obtained 68,588 tourists to Tokyo, 706,221 tourism-related tweets, and 218,052 tourism period documents. To prepare a ground truth of classification, we annotate tourism labels ("sightseeing", "business", "eat", "buy", and "other") to tourism tweets, and purpose labels ("sightseeing", "business", and "other"). In this process, the author annotates labels to tourism tweets and tourism period documents based on the contents to fit their natural feelings. Table 2 presents the number of each labels.

Table 2. Number of classes and labels of tourism tweets and tourism period documents

	sightseeing	business	eat	buy	other	total
Tourism tweet	1,073	702	928	659	1,754	5,107
Tourism period document	381	468			349	1,198

We vectorize the tourism period documents and tourism tweets using the process described in Sect. 3.3. The dimensions are increased from 100 to 500 dimensions by 100 dimension steps. Based on experiments conducted with multi-class SVM, we used 400-dimensional in following experiments, which scores the highest accuracy.

All SVMs employed for the experiments used a Gaussian kernel. Additionally, we experiment using five cross-validation, and hyperparameters c and γ, which have the highest accuracy in each experiment. We used Accuracy, Precision, Recall, and the F-Score as evaluation criteria to compare classification results.

4.2   Experimental Procedures

This section presents a description of the method of the experiments performed in this paper. First, Fig. 1 presents an overall view of the experimental.

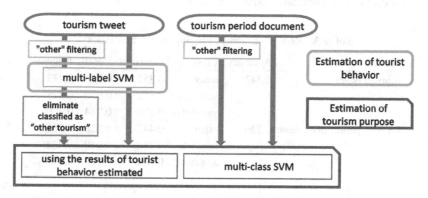

Fig. 1. Overview of the experiment

For evaluation of tourist behavior inferences, we experiment with two methods. First, we apply "other" filtering to tourism tweets, and infer tourist behavior by a multi-label SVM. It is named "filtering and SVM". Second, we do not apply "other" filtering to tourism tweets, and infer tourist behavior. It is named "only SVM".

For tourism purpose inference, we experiment with four methods. First method is that we infer the tourism purpose using the inferred class of tourist behavior in "filtering and SVM". At the time, we eliminate the tweets classified as "other tourism". It is named "eliminate and using result of SVM". Second method is inference of tourism purposes using all tourism tweets (including "other tourism"). It is named "only using result of SVM". Third method is applied to "other" filtering to the tourism period documents. We infer tourism purposes using a multi-class SVM. It is named "filtering and using document". In addition, fourth method is one by which we infer tourism purpose using a multi-class SVM without a procedure to eliminate "other" filtering from tourism period documents. It is named "only using document".

4.3   Experiment Results and Discussion

4.3.1   Evaluation of "Other" Filtering

First, we evaluate "other" filtering to eliminate "other tourism" and "other purpose". Our method uses "other" filtering in "filtering and SVM", to tourism tweets, and in "filtering and using document", to tourism period documents. Table 3 portrays the classification results of "other" filtering for tourism tweets and tourism period documents. Table 4 shows the number of tourism tweets and tourism period documents erased by "other" filtering. Here, Accuracy, Precision, Recall, and F-Score show average values in each class. In Table 3, recall of "other" is low in both the tourism tweets and tourism period documents. Moreover, in Table 4, the most eliminated class is "other" of tourism tweets and tourism period document about 33%. In addition, the other class accounts for a few percent. Although this indicates that eliminating "other" is insufficient, this procedure is effective for "other behavior" and "other purpose" because the recall of "non-other" is high (i.e., misclassification of those classes is less, but this procedure can eliminate "other" classes).

Table 3. Results of "other" filtering: evaluation index

Applies	Accuracy	Class	Precision	Recall	F-score
Tourism tweet	0.742	other	0.852	0.326	0.472
		non-other	0.722	0.970	0.828
		avg./total	0.770	0.744	0.700
Tourism period document	0.730	other	0.442	0.228	0.266
		non-other	0.752	0.940	0.834
		avg./total	0.664	0.730	0.670

Table 4. Results of "other" filtering: number of erasures

Applies	Class	Before	After	Erasure rate (%)
Tourism tweet	business	702	665	5.270
	sightseeing	1,073	1,038	3.262
	buy	659	643	2.428
	eat	928	917	1.185
	other	1,745	1,177	32.550
Tourism period document	business	468	447	4.487
	sightseeing	381	351	7.874
	other	349	269	22.922

The reason why the recall of "other" is low is the diversity of tweets. Results show that the contents of those tweets are various. It is unrealistic to train the characteristics of those "other" tweets to eliminate "other" classes. Therefore, we infer that "other" filtering is more important to classify "other" classes mistakenly into other classes.

4.3.2   Evaluation of Tourist Behavior Inference

We describe the performance of tourist behavior inference. For tourist behavior inference, two types of the experimental methods are shown in Sect. 4.2. The experimentally obtained results for each of the method are shown in Table 5. There, the average F-score of all classes in "filtering and SVM" and "only SVM" is about 0.65.

The F-score of "sightseeing" in "filtering and SVM" and "only SVM" is low, compared to each attribute. Here, from Table 1, many types of behavior corresponding to the "sightseeing" than "business". Therefore, "sightseeing" is more diverse than the others and the vector of the tourism tweets is a variation to classify. As a result, the F-score of the "sightseeing" in classification is regarded as lower than other attributes.

Table 5. Classification results of tourist behavior

Method	Accuracy	Label	Precision	Recall	F-score
Filtering and SVM	0.544	business	0.758	0.646	0.698
		sightseeing	0.642	0.556	0.596
		buy	0.788	0.666	0.722
		eat	0.744	0.652	0.694
		other	0.616	0.572	0.592
		avg./total	0.698	0.612	0.648
Only SVM	0.570	business	0.718	0.650	0.676
		sightseeing	0.638	0.536	0.582
		buy	0.724	0.676	0.696
		eat	0.728	0.644	0.684
		other	0.710	0.706	0.710
		avg./total	0.704	0.646	0.672

Tables 6 and 7 present results and the labels of ground truth of "filtering and SVM" and of "only SVM". In those tables, "correct" represents the labels of ground truth, which were annotated manually. "estimate" represent the labels inferred using the proposed method. In those tables, almost all tourism tweets of "buy" were classified correctly because many words such as "buy" or "souvenirs", which are clearly related to buying behavior, appear in the text of the tourism tweet of the class "buy". In addition, some tourism tweets related to "sightseeing" were misclassified as "other behavior". We consider that vector variation of tourism tweets can lead to responses of "other behavior" and "sightseeing", as described earlier. The class of "sightseeing" is divided to sub-classes that contain more details of behavior.

Furthermore, in Tables 6 and 7, some tweets are classified into "unlabeled". The multi-label SVM is configured with multiple two-class SVMs in each attribute. The SVM classifies a tourism tweet according to its attributes or not. Therefore, the SVMs do not often classify the tweet into any class. However, the tweet is originally classified as "business". This method misclassifies the tweet as "unlabeled".

Here, we discuss that we apply "other" filtering to tourism tweets or not. In Table 5, the average of each criterion of "filtering and SVM" is less than "only SVM". In addition, the value of "other tourism" in "filtering and SVM" is less than "only SVM", although "filtering and SVM" eliminates tourism tweets "other tourism". The evaluation value of the classes other than "other tourism" in "filtering and SVM" is higher than "only SVM". This result presents the performance of "other" filtering is sufficient, but "other tourism: in "filtering and SVM" is difficult to eliminate because classifying easily the tourism tweets is already finished. In addition, the reason why the evaluation value of "other tourism" in "filtering and SVM" is less than "only SVM" is the average contains score of classifying tourism tweets into "other tourism" or not. However, because our purpose is inferring the tourist behavior, we regard the evaluation values of classes other than "other tourism" are more important. Therefore, we regard "filtering and SVM" as better than "only SVM" to infer tourist behavior.

Table 6. Results of labeling using "filtering and SVM"

correct\estimate	business	sightseeing	buy	eat	other	unlabeled
business	432	53	16	70	88	79
sightseeing	42	579	52	78	206	155
buy	19	42	430	49	49	90
eat	71	89	44	598	85	113
other	78	175	42	81	671	212

Table 7. Results of labeling using "only SVM"

correct\estimate	business	sightseeing	buy	eat	other	unlabeled
business	452	47	25	79	115	82
sightseeing	48	577	57	76	232	169
buy	23	49	444	52	68	94
eat	85	82	49	597	105	127
other	64	186	76	92	1,230	213

4.3.3 Evaluation of Tourism Purpose Inference

We describe the performance of the tourism purpose inference. In Table 8, we present results of tourism purpose inference performance. This evaluation compares 4 methods. Compared to each method, it is apparent that classification performance using the tourist behavior is better than that of multi-class SVM using the tourism period documents. In the proposed method, although we regard all tweets during the tourism period as tourism period documents, the key tweets used to ascertain the class of tourism purpose are few tweets during the tourism period document. Therefore, tweets other than the key tweets might adversely affect classification performance. As a result, "eliminate and using result of SVM" and "only using result of SVM" are better than "filtering and using document" and "only using document" to infer tourism purposes.

Next, comparison of methods reveals that the tourism purpose inference performance is improved by eliminating "other purpose" because the classification performance of the other class is improved by erasing "other purpose". In addition, a tourism period document that does not contain "sightseeing" and "business" during the tourism period is not classified as "other purpose". In Sect. 5.2, we confirmed that the classification performance of other attributes improves by eliminating the previous "other".

Table 8. Classification results of tourism purposes

Method	Accuracy	Class	Precision	Recall	F-score
Eliminate and using result of SVM	0.760	business	0.830	0.770	0.800
		sightseeing	0.740	0.720	0.730
		other	0.700	0.780	0.740
		avg./total	0.760	0.760	0.760

(*continued*)

Table 8. (*continued*)

Method	Accuracy	Class	Precision	Recall	F-score
Only using result of SVM	0.720	business	0.790	0.740	0.760
		sightseeing	0.750	0.640	0.690
		other	0.620	0.770	0.690
		avg./total	0.730	0.720	0.720
Filtering and using document	0.654	business	0.704	0.796	0.744
		sightseeing	0.664	0.640	0.650
		other	0.554	0.452	0.482
		avg./total	0.654	0.654	0.646
Only using document	0.652	business	0.654	0.852	0.742
		sightseeing	0.682	0.546	0.602
		other	0.634	0.504	0.556
		avg./total	0.658	0.654	0.642

However, in the tourist behavior, it is regarded as the "other purpose" is a set of "other behavior", "eat", and "buy". Therefore, the classification performance of "other purpose" is considered to have improved in the tourism purpose. Consequently, it is probably useful to erase "other purpose" for tourism purpose inference.

5 Conclusion

For this research, we proposed a method to infer user tourism purposes and tourist behaviors at a tourist destination using tweet geo-tagging and text posted to Twitter as one approach to extract tourist information. Our proposed method classifies a feature vector by a multi-label SVM as tourist behavior related to "sightseeing", "business", "eat", "buy" and "other behavior". Subsequently, we classify tourism period documents into tourism purposes of "tourism", "business", and "other" using tourist behavior inference results. The evaluation experiment showed F-scores of tourist behavior and tourism purposes were both about 0.65.

As future applications, we expect to visualize classification results obtained using the proposed method based on time. The obtained valid information might increase tourist arrivals by enabling them to visualize the area, and to analyze tourist behavior and position at the time of that behavior according to tourism purposes.

Acknowledgements. This work was supported by JSPS KAKENHI Grant Number 16K00157, 16K16158, and Tokyo Metropolitan University Grant-in-Aid for Research on Priority Areas "Research on social big data". We are grateful for the assistance by Yoshiyuki Shoji.

Refernces

1. Fang, G., Sayaka, K., Satoshi, F.: How to extract seasonal features of sightseeing spots from Twitter and Wikipedia (Preliminary Version). Bull. Network. Comput. Syst. Softw. **4**(1), 21–26 (2015)
2. Bannur, S., Omar, A.: Analyzing temporal characteristics of check-in data. In: Proceedings of the Companion Publication of the 23rd International Conference on World Wide Web, pp. 827–832 (2014)
3. Burger, J.D., Henderson, J., George, K., Zarrella, G.: Discriminating gender on Twitter. In: Proceedings of the Conference on Empirical Methods in Natural Language Processing, pp. 1301–1309 (2011)
4. Rao, D., Yarowsky, D., Shreevats, A., Gupta, M.: Classifying latent user attributes in Twitter. In: Proceedings of the Second International Workshop on Search and Mining User-Generated Contents, pp. 37–44 (2010)
5. Burger, J.D., John C.H.: An exploration of observable features related to blogger age. In: AAAI Spring Symposium: Computational Approaches to Analyzing Weblogs, pp. 15–20 (2006)
6. Pennacchiotti, M., Ana M.P.: Democrats, republicans and starbucks aficionados: user classification in Twitter. In: Proceedings of the 17th ACM SIGKDD International Conference on Knowledge Discovery and Data Mining, pp. 430–438 (2011)
7. Cheng, Z., James, C., Kyumin, L.: You are where you tweet: a content-based approach to geo-locating Twitter users. In: Proceedings of the 19th ACM International Conference on Information and Knowledge Management, pp. 759–768 (2010)
8. Sloan, L., Morgan, J., Burnap, P., Williams, M.: Who tweets? Deriving the demographic characteristics of age, occupation and social class from Twitter user meta-data. PLoS ONE **10**(3), e0115545 (2015)
9. Li, Y.R., Wang, Y.Y.: Exploring the destination image of chinese tourists to taiwan by word-of-mouth on web. In: Proceedings of World Academy of Science, Engineering and Technology, No. 79, p. 977 (2013)
10. Al Zamal, F., Wendy, L., Derek, R.: Homophily and latent attribute inference: inferring latent attributes of twitter users from neighbors. In: International Conference on Weblogs and Social Media, vol. 270 (2012)
11. Ishino, A., Nanba, H., Takezawa, T.: Automatic compilation of an online travel portal from automatically extracted travel blog entries. In: ENTER, pp. 113–124 (2011)
12. Vapnik, V.N., Vlamimir, V.: Statistical Learning Theory, vol. 1 (1998)
13. Benevenuto, F., Magno, G., Rodrigues, T., Almeida, V.: Detecting spammers on Twitter. In: Collaboration, Electronic Messaging, Anti-Abuse AND Spam Conference, vol. 6, p. 12 (2010)
14. Pennacchiotti, M., Ana-Maria, P.: A machine learning approach to Twitter user classification. In: International Conference on Weblogs and Social Media, vol. 11. No. 1, pp. 281–288 (2011)
15. Deerwester, S., Dumais, S.T., Furnas, G.W., Landauer, T.K., Harshman, R.: Indexing by latent semantic analysis. J. Am. Soc. Inf. Sci. **41**(6), 391 (1990)

IWEC 2016: 7th International Workshop on Empathic Computing

Application of Annotation Smoothing
for Subject-Independent Emotion Recognition
Based on Electroencephalogram

Nattapong Thammasan[1(✉)], Ken-ichi Fukui[2], and Masayuki Numao[2]

[1] Graduate School of Information Science and Technology,
Osaka University Suita-Shi, Osaka, 565-0871, Japan
nattapong@ai.sanken.osaka-u.ac.jp
[2] Institute of Scientific and Industrial Research (ISIR), Osaka University Ibaraki-Shi,
Osaka, 567-0047, Japan
{fukui,numao}@ai.sanken.osaka-u.ac.jp

Abstract. In the construction of computational models to recognize emotional state, emotion reporting continuously in time is essential based on the assumption that emotional responses of a human to certain stimuli could vary over time. However, currently existing methods to annotate emotion in temporal continuous fashion are confronting various types of challenges. Therefore, the manipulation of the annotated emotion prior to labeling training samples is necessary. In this work, we present an early attempt to manipulate the emotion annotated in arousal-valence space by applying three different signal filtering techniques to smooth annotation data; moving average filter, Savitzky-Golay filter, and me-dian filter. We conducted experiments of emotion recognition in music listening tasks employing brainwave signals recorded from an electroencephalogram (EEG). Smoothed annotation data were used to label the features extracted from EEG signals to train emotion recognizers using classification and regression techni-ques. Our empirical results indicated the potential of the moving average filter that could increase the performance of emotion recognition evaluated in subject-independent fashion.

Keywords: Emotion recognition · Electroencephalogram · Music-emotion · Annotation · Smoothing

1 Introduction

Among the endeavors of building computational models capable of perceiving human emotion, recent researchers have emphasized the necessity of estimating emotional state continuously over the course of time (Gunes and Schuller 2013). In particular, the empathic computing systems are expected to be capable of capturing emotional fluctu-ation of users and properly respond almost instantly. Beyond the continuity in time, psychological researchers proposed to describe emotion in multi-dimensional contin-uous space as a dimensional continuous model could resolve the ambiguity issues occurred in using discrete categories to describe emotion. Among these, arousal-valence

© Springer International Publishing AG 2017
M. Numao et al. (Eds.): PRICAI 2016 Workshops, LNAI 10004, pp. 115–126, 2017.
DOI: 10.1007/978-3-319-60675-0_10

model (Russell 1980) was one of the most commonly exploited models to describe emotional states; arousal indicates emotional intensity ranging from calm to activated emotion, whereas valence, on the other dimension, describes positivity of emotion ranging from unpleasant to pleasant. Considering the combined continuity of emotion in time and space, the efforts to track emotion variation in the continuous arousal-valence space have emerged.

However, the approach is still in its infancy and confronting with a variety of challenges. In particular, it lacks standard methodology to report emotion continuously in time and space. The existing emotion annotation approaches have certain limitation (Metallinou and Narayanan 2013). For instance, unfamiliarity with annotation tools could lead to the inconsistency and the contamination of noise in emotion annotation data while familiarizing the user with annotation tools by increasing practicing time could be annoying and lead to the imprac- ticality of the systems. In addition, the delay of annotation could result in a mismatch between the emotional cues and the reported emotions (Mariooryad and Busso 2013). Therefore, completely relying on the annotation data reported by the user would degrade the performance of emotion recognition systems. By these reasons, the manipulation of emotion annotation data prior to proceeding to emotion recognition system construction is undoubtedly essential but was usually not taken into consideration in previous research.

In this work, we propose an early attempt of manipulating user's emotion annotation data continuously self-reported through time in the continuous arousal- valence space. In particular, we introduce an application of signal smoothing techniques to alleviate the adverse effect of annotation noises. While continuous emotion annotation can be perform using a variety tools (e.g., mouse- moving (Cowie et al. 2000) and joystick (Soleymani et al. 2012)), we focus on mouse-clicking emotion annotation approach for the sake of simplicity and less cognitive processing requisition. In our experiment, electroencephalogram (EEG) was used to record brainwave signals in music listening tasks, then informative features were extracted from EEG responses and such features were labeled with the smoothed annotation data. Afterward, the continuous emotion recognition models were established by machine learning approaches using classification technique (Thammasan et al. 2016) and regression technique (Soleymani et al. 2016). The performance of emotion recognition was evaluated by using subject-independent approach aiming the border goal toward generic emotion recognition systems.

2   Research Methodology

2.1   Experimental Protocol

Fifteen healthy male subjects aged 22–30 years (mean age = 25.52 years, SD = 2.14 years) participated in the experiment. All of them were students of Osaka University and had a minimal formal musical education. At the beginning, each subject was instructed to select 16 MIDI songs from a 40-song music collection (mean length = 106.3 s, SD = 16.2 s, range = 73-147 s). For further investigation the selected songs were controlled to be comprising of an equal number of familiar and unfamiliar songs.

Afterward, the selected songs were presented to the subject as synthesized sounds using the Java Sound APIs MIDI package[3]; a 16 s silent resting period was inserted at the interval between each song to reduce any effect of the previous song. During music listening, brainwave signals were recorded using Waveguard EEG[4] placed in accordance with the 10–20 international system, and acquired at a sampling rate of 250 Hz. Positions of the twelve selected electrodes (Fpl, Fp2, F3, F4, F7, F8, Fz, C3, C4, T3, T4, and Pz) were nearby frontal lobe, which is believed to be play an importance role in emotion processing (Koelsch 2014), whereas the vertex electrode (Cz) served as a reference electrode. Over the entire course of EEG recording, the impedance of each electrode was maintained below 20 kΩ. EEG signals were amplified by Polymate AP1532 amplifier[5] and visualized on APMonitor[6]. Each subject was instructed to keep his eyes close and limit body movement during music listening to alleviate any effect owing to unrelated artifacts. The EEG signals were filtered using a 60-Hz notch filter to suppress the noise of the electric power line followed by a 0.5–60 Hz band-pass filter to eliminate the unrelated noises. Eye-movement artifacts were corrected by applying the independent component analysis of the EEGLAB toolbox (Delorme et al. 2011).

After finishing the listening session of all sixteen songs, each subject proceeded to the emotion annotation session without EEG recording. The annotation was conducted through our developed software. Each subject was instructed to annotate the emotions that were perceived in the previous session by continuously specifying at a corresponding point on the arousal-valence space displayed on a monitor screen (Fig. 1). At the end of the emotion annotation of each song, each subject rated the confidence level, on a scale of ranging from 1 to 3, of the correspondence between the emotions perceived during the first listening phase and the annotated emotions. Each subject was also encouraged to perform the emotion annotation of a particular song again in the case that the annotated data for that song was not satisfied yet. A brief guideline of arousal-valence emotion model was provided throughout annotation session to acquaint each subject

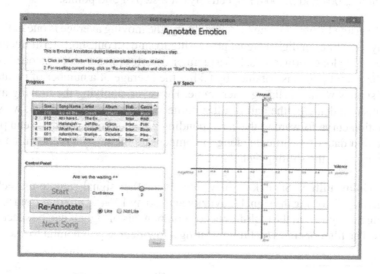

Fig. 1. A screenshot of the annotation software

with the model. Arousal and valence annotation data were recorded independently as numerical values ranging from -1 to 1. Eventually, the annotation data was associated with the artifact-corrected EEG signals via timestamps. Unfortunately, we discarded all data from the two subjects who reported drowsiness during EEG recording.

2.2   Annotation Smoothing

The assumption for the mouse-clicking emotion annotation is that annotated emotion is stable for a certain duration until the next annotation is clicked.

However, this assumption is quite far from realistic because emotion in real life gradually changes from one to another state rather than abruptly shifts from one to another. Therefore, our mouse-clicking emotion annotation tool has limited practicability despite our encouragement for subject to annotate emotion in a high frequency. Consequently, adjusting the curve of emotion annotation that could provide a smoother curve reflecting the gradual changing of emotion might be necessary as the curve could be considered as more corresponding to the real characteristic of emotion. In addition, the noises owing to unfamiliarity to the tools of some annotators and unintentionally mouse clicking could inevitably contaminate the annotation data. Therefore, adjusting the annotation curve by using a certain approximation function could also correct these noises.

To resolve the issues, we hereby propose the techniques to manipulate the an-notation data by applying time-series data smoothing techniques directly to the annotation curves. Inspired by signal processing techniques, we introduce an application of three commonly used filtering techniques implemented by MATLAB Signal Processing Toolbox[1] to smooth our annotation curve. As the characteristic of the smoothed curves highly relies on the size of filter frame, we also examined the influence of the size to our emotion recognition performance by varying the filter frame size from 501 to 8001 points (equivalent to 2.004 and 32.004, respectively) at a step of 500 points.

Moving Average Filter.   Owing to its simplicity, the moving average is one of the most commonly used filter in digital signal processing to smooth out short- term fluctuations while preserving long-term trends. It is a premier filter for time domain encoded signals. Each of the output points is obtained by taking the average of a number of points from the input signals within a sliding filter frame. However, applying the moving average filter results in a delay by the half of filter frame size. To handle the delay, we shifted the annotation curve forward to the corresponding point and replicated the end point in the annotation data of a particular song to compensate the missing data with a size of the delay owing to the shifting technique.

Savitzky-Golay Filter.   Savitzky-Golay filter is another signal smoothing technique without greatly distorting the signal (Savitzky and Golay 1964). Different from applying the moving average filter, the high-frequency components of signals can be successfully preserved. The filter operates by fitting a fitting a low-degree polynomial to a set of data

---

[1]   http://www.mathworks.com/products/signal/.

points in a sliding filter frame by the method of local least-squares polynomial approximation and then evaluating the resulting polynomial at a single point within the approximation interval. In this work, we applied Savitzky-Golay filter to smooth annotation data with the two different levels of polynomial degree—three (cubic) and four (quartic) —to examine the effect of the polynomial degree to the smoothed annotation data and the performance of the following emotion recognition.

Median Filter. Unlike the other filters that mainly eliminate the sharp edges of input signals, a median filter is an alternative filter that has prominence in smoothing input signals while preserving the edge of signals. Within a sliding filter frame, smoothed signals are obtained by deriving the median of a number of points from the input signals.

2.3 Experiments of Emotion Recognition

Inspired by the successful results in previous EEG-based emotion recognition studies (Sourina et al. 2012; Thammasan et al. 2016), the fractal dimension (FD) approach was exploited to extract informative features from the EEG signals. FD value is a non-negative real value that characterizes the complexity and irregularity of a time-varying signal and it could be used to indicate brain states from EEG signals (Sourina et al. 2011). In this study, we implemented Higuchi algorithm (Higuchi 1988) to calculate FD values. As reported as effective features to estimate emotional states (Koelstra et al. 2012; Thammasan et al. 2016; Lin et al. 2014), asymmetric features were also included in our original feature set; the feature was derived from the difference of FD value extracted from an electrode on the left hemisphere and that from the symmetric lateral electrode on the right hemisphere. As there were five symmetric electrode pairs, the total number of the features extracted from EEG signals is, therefore, 17. To capture temporal dynamics of the emotional states, we applied a non-overlapping 4 s sliding window segmentation technique; the size of 3–6 s was reported to achieve the highest performance in emotion classification in literature (Candra et al. 2015).

Afterward, the extracted features were labeled by the corresponding annotation data prior to constructing emotion recognition models.

Next, we conducted the experiments of emotion recognition by applying classification and regression techniques. For the sake of simplicity, the outputs of emotion classification are the binary classes of arousal and valence, while the outputs of regression are numerical numbers ranging from −1 to 1 of arousal and valence estimation in the two-dimensional emotion space.

Classification. Despite the continuity of the arousal-valence space, we converted emotion recognition into the binary classification of arousal and valence independently. Arousal classification was to classify high and low arousal, while valence classification was to classify positive and negative valence, whereas the sign of the annotated numerical rating was used to define the class of arousal and valence. A majority method was adopted to determine the emotional label for a particular window containing emotional class shifting. The features were scaled for each subject between [0, 1] using *min-max* strategy. As a classifier, we adopted support vector machine (SVM) based on Gaussian radial basis kernel function (RBF) using MATLAB Statistics and Machine Learning Toolbox[8]; the SVM was found to be popular and successful classifier int the research of EEG- based emotion recognition (Kim et al. 2013). The kernel scale of RBF kernel was set as 0.5.

The evaluation of emotion classification was performed by using subject- independent approach. This approach trains and tests emotion classification model using the aggregated data from all subjects. It can apparently reflect the degree of generalization of the emotion recognition model. In this work, we adopted the leave-one-subject-out validation method to evaluate the performance of classification. In each trial, the classifier was trained by using the combined data from twelve subjects and then the trained classifying model was tested against the data from the remaining subject. The results from each trial were averaged to derive overall performance.

Regression. Utilizing the continuity of arousal-valence space, we also performed emotion recognition by using regression technique to recognize arousal and valence independently as the technique could provide an estimated emotion as numerical values of arousal and valence in the arousal-valence space. Compared to classification, regression could provide finer detail of the estimated emotion and the different emotions belonging to the same quadrant in the arousal-valence space can be distinguished. In this work, we applied the support vector machine regression based on Gaussian kernel implemented by using MATLAB Statistics and Machine Learning Toolbox[3] was used to estimate emotion; the kernel scale of Gaussian kernel was set as 0.5. To label the features extracted from a particular sliding window with an emotional tag, the averaged values of arousal and valence in that window were used.

The performance of emotion recognition using regression was evaluated by using the Pearson correlation to reflect the similarity between the estimated curves and the annotation data, and the mean square error (MSE) to represent the disparity (loss) between the estimated emotion and the ground truth. Similar to emotion classification, the overall performance of emotion recognition was evaluated in subject-independent strategy by using leave-one-subject-out cross- validation.

3   Results

For clarity, the main purpose of this current work is to study the feasibility of applying the smoothing technique to the annotation data aiming to enhance the performance of emotion recognition. First, we examined the shapes of the smoothed annotation curves. Next, we assessed the performance of emotion recognition with the smoothed annotation data using subject-independent evaluation. The high averaged confidence level (2.4109, SD = 0.6676) of the annotation across subjects suggested the validity of the annotation data.

3.1 Annotation Smoothing Results

To illustrate the results of applying the smoothing technique to the annotation data, we exemplify by using the emotion annotation data of a song by Subject 12. We varied the size of filter frame from small (501 points) to medium (4001 points) and large (8001 points) frame. The comparison between original annotation data, which contains the variation of the arousal and valence over the course of time, and the smoothed annotation data is shown in Fig. 2. As can be seen, the smoothed annotation data was similar to the original annotation data when using a small filter frame size. Specifically, the moving average filter could noticeably remove the high-frequency fluctuation in the annotation. However, when the filter frame size was enlarged to medium level, we found that the trend of annotation curve was preserved but the high-frequency fluctuation was dramatically reduced for any filtering technique. Interestingly, flat plateaus were also found in the resulted annotation data when applying the median filter. Nevertheless, when we applied filtering technique with large filter frame size, the shape of the annotation data was highly distorted in comparison to the original annotation data, especially for the moving average filter and the median filter. Specifically, the height of the resulted curve was distinctly dissimilar to the original curve when applying either filter, while the Savitzky-Golay filter could still preserve the height of the curve regardless of the polynomial order.

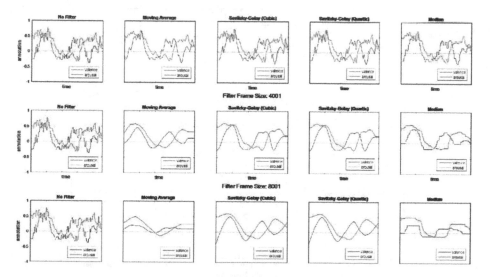

Fig. 2. An example of the resulting smoothed annotation data compared to the original annotation data of a song by Subject 12 (song length = 91.6 s)

3.2 Results Using Classification

The averaged classification accuracies across subjects are illustrated in Fig. 3. As can be seen, the obtained accuracies were higher than the random classification (50% accuracies for two-class classification) but the limited performance suggested that the

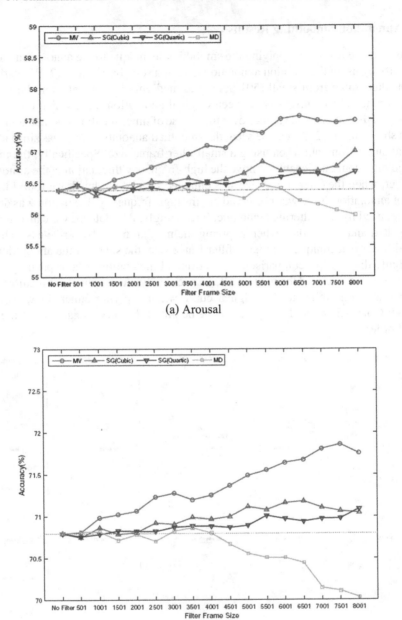

(a) Arousal

(b) Valence

Fig. 3. Results of subject-independent emotion classification using the annotation data smoothed by the moving average (MV), Savitzky-Golay (SG), and median (MD) filters

classification might suffer from the inter-subject variability in EEG signals and/or in emotion annotation strategies. The enhancement of the emotion classification perform-ance as filter frame size enlarged was found, especially by using the moving average filter. The Savitzky-Golay filter also achieved the improved classification results, where the cubic filter could outperform the quartic filter. In overall, the results suggested the promise of the moving average filter to upgrade the performance of emotion recognition.

3.3 Results Using Regression

The performance of emotion recognition using regression is shown in Fig. 4. Similar to emotion classification, the low performance of the subject-independent emotion recognition suggested the adverse effect of the inter-subject variability. Despite the limited performance, all of the approaches could enhance the performance of arousal estimation, especially when increasing the filter frame size to a certain extent. Moving average filter achieved the improved performance in majority cases of arousal and valence recognition suggesting the promise of this technique. The proper size of filter frame was approximately between 18 s (4501 points) and 24 s (6001 points). In addition, the median filter was found to be another potential approach to upgrade arousal recognition performance. Furthermore, the lower polynomial order of Savitzky-Golay filter can be preferred as the cubic filter achieved better performance in comparison to the quartic filter.

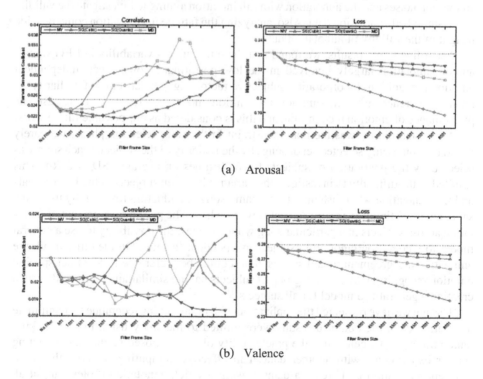

(a) Arousal

(b) Valence

Fig. 4. Performance of subject-independent emotion recognition using regression evaluated on the annotation data smoothed by the moving average (MV), Savitzky-Golay (SG), and median (MD) filters

4 Discussion

The primary goal of this current work is to improve the performance of emotion recognition, and we propose techniques of smoothing annotation data as an approach to manipulate the annotation data prior to training emotion recognizer. The empirical results suggested that our proposed methods could slightly enhance the performance of emotion recognition either using classification or regression techniques. However, several issues leave room for discussion.

Even though the moving average filter has demonstrated its promise in improving emotion recognition, the considerable distortion of the smoothed annotation data in comparison to the original data (shown in Fig. 2) suggested the necessity of trading off between enhancing the performance of emotion recognition and the disparity between the smoothed annotation curve and the original curve. With the consideration of sufficient accuracy of the smoothed annotation data, a statistical parameter capable of indicating the extent of such distortion is encouraged to be introduced in future works. In addition, future study should also include the reporting of annotation strategy and annotation fatigue of each annotator to enable the feasibility to distinguish the unintentional annotation noises and the annotation with full intention aiming to investigate the validity of the smoothed annotation data. Also, analyzing the fatigue in annotation could provide insight of the validity of annotation data.

According to the results, we found that the inter-subject variability in EEG signals and/or annotation largely involved in the low performance of subject- independent emotion recognition. Incorporating subjective factor, e.g. gender, music familiarity, or music preference, into the emotion recognition model is expected to increase the performance of emotion recognition, and this is considered as our possible future work.

As the familiarity was the main constraint in the song selection phase, we barely found the commonly selected set of songs by the majority of the subjects; each song was selected by the population of subjects for 6.00 times on average (SD = 2.76). This resulted in the difficulty to investigate the variance between subjects of the EEG signals and the annotation when listening to the same song. Conducting research by using the same musical stimuli for every subject is worthwhile. The possible agreement of annotation across subjects in a particular song would suggest the possibility to use a general model to recognition emotion in subject-independent fashion. On the other hand, the annotation disagreement would suggest that it could be preferable to construct specific emotion recognition model for a group of subjects having similar annotation rather than creating a generalized model for all subjects.

Though we have presented the application of data smoothing technique to manipulate annotation data, the obtained results were limited to the study using mouse-clicking annotation tool. To generalize the practicability of our proposed methods, conducting further experiments with another dataset in affective computing research that used different annotation tools (e.g., a dataset using joystick annotation (Soleymani et al. 2012)) is also encouraging. In addition, with the aim to generalize the applicability of our emotion recognition approach, our future works include conducting experiments with broader groups of subjects; for instance, recruiting female subjects or aging population to participate in our research could be done in the future.

5 Conclusion

In this work, we introduce the methods to manipulate the emotion annotation data prior to proceeding into emotion recognition phase. We applied signal smoothing techniques directly to the acquired annotation data. Based on the empirical results, evaluated in subject-independent fashion, of EEG-based arousal and valence recognition, the moving average filter demonstrated its promise in enhancing the performance of emotion classification and tracking. However, the issue of annotation curve distortion owing to smoothing approaches is a subject to be studied in the future work.

Acknowledgment. This research is partially supported by the Center of Innovation Program from Japan Science and Technology Agency (JST), JSPS KAKENHI Grant Number 25540101, and the Management Expenses Grants for National Universities Corporations from the Ministry of Education, Culture, Sports, Science and Technology of Japan (MEXT).

References

Candra, H., Yuwono, M., Chai, R., Handojoseno, A., Elamvazuthi, I., Nguyen, H., Su, S.: Investigation of window size in classification of EEG-emotion signal with wavelet entropy and support vector machine. In: Proceedings of the 37th Annual International Conference of the IEEE Engineering in Medicine and Biology Society, pp. 7250–7253 (2015)

Cowie, R., Douglas-Cowie, E., Savvidou, S., McMahon, E., Sawey, M., Schröder, M.: Feeltrace: an instrument for recording perceived emotion in real time. In: Proceedings of the ISCA Workshop on Speech and Emotion, pp. 19–24 (2000)

Delorme, A., Mullen, T., Kothe, C., Acar, Z.A., Bigdely-Shamlo, N., Vankov, A., Makeig, S.: EEGLAB, SIFT, NFT, BCILAB, and ERICA: new tools for advanced EEG processing. Comp. Intell. Neurosci. (2011)

Gunes, H., Schuller, B.: Categorical and dimensional affect analysis in continuous input: Current trends and future directions. Image Vis. Comput. **31**(2), 120–136 (2013)

Higuchi, T.: Approach to an irregular time series on the basis of the fractal theory. Physica D **31**(2), 277–283 (1988)

Kim, M.K., Kim, M., Oh, E., Kim, S.P.: A review on the computational methods for emotional state estimation from the human EEG. Comput. Math., Methods in Medicine (2013)

Koelsch, S.: Brain correlates of music-evoked emotions. Nat. Rev. Neurosci. **15**(3), 170–180 (2014)

Koelstra, S., Muhl, C., Soleymani, M., Lee, J.S., Yazdani, A., Ebrahimi, T., Pun, T., Nijholt, A., Patras, I.: DEAP: a database for emotion analysis using physiological signals. IEEE Trans. Affect. Comput. **3**(1), 18–31 (2012)

Lin, Y.P., Yang, Y.H., Jung, T.P.: Fusion of electroencephalogram dynamics and musical contents for estimating emotional responses in music listening. Front. Neurosci. 8(94) (2014)

Mariooryad, S., Busso, C.: Analysis and compensation of the reaction lag of evaluators in continuous emotional annotations. In: Proceedings of the 5th Humaine Association Conference on Affective Computing and Intelligent Interaction, pp. 85–90 (2013)

Metallinou, A., Narayanan, S.: Annotation and processing of continuous emotional attributes: Challenges and opportunities. In: Proceedings of the 10th IEEE International Conference and Workshops on Automatic Face and Gesture Recognition. pp. 1–8 (2013)

Russell, J.A.: A circumplex model of affect. J. Pers. Soc. Psychol. **39**(6), 1161–1178 (1980)

Savitzky, A., Golay, M.J.E.: Smoothing and differentiation of data by simplified least squares procedures. Anal. Chem. **36**(8), 1627–1639 (1964)

Soleymani, M., Asghari-Esfeden, S., Fu, Y., Pantic, M.: Analysis of eeg signals and facial expressions for continuous emotion detection. IEEE Trans. Affect. Comput. **7**(1), 17–28 (2016)

Soleymani, M., Lichtenauer, J., Pun, T., Pantic, M.: A multimodal database for affect recognition and implicit tagging. IEEE Trans. Affect. Comput. **3**(1), 42–55 (2012)

Sourina, O., Liu, Y., Nguyen, M.K.: Real-time EEG-based emotion recognition for music therapy. J. Multimodal. User. Int. **5**(1–2), 27–35 (2012)

Sourina, O., Wang, Q., Liu, Y., Nguyen, M.K.: A real-time fractal-based brain state recognition from eeg and its applications. In: Babiloni, F., Fred, A.L.N., Filipe, J., Gamboa, H. (eds.) Proceedings of the BIOSIGNALS, pp. 82–90 (2011)

Thammasan, N., Moriyama, K., Fukui, K., Numao, M.: Continuous music- emotion recognition based on electroencephalogram. IEICE Trans. Inform. Syst. **E99-D**(4), 1234–1241 (2016)

Modeling Negative Affect Detector of Novice Programming Students Using Keyboard Dynamics and Mouse Behavior

Larry Vea[1,2(✉)] and Ma. Mercedes Rodrigo[2]

[1] Mapua Institute of Technology, Makati City, Philippines
lavea@mapua.edu.ph
[2] Ateneo de Manila University, Quezon City, Philippines
mrodrigo@ateneo.edu

Abstract. We developed affective models for detecting negative affective states, particularly boredom, confusion, and frustration, among novice programming students learning C++, using keyboard dynamics and/or mouse behavior. The keystroke dynamics are already sufficient to model negative affect detector. However, adding mouse behavior, specifically the distance it travelled along the x-axis, slightly improved the model's performance. The idle time and typing error are the most notable features that predominantly influence the detection of negative affect. The idle time has the greatest influence in detecting high and fair boredom, while typing error comes before the idle time for low boredom. Conversely, typing error has the highest influence in detecting high and fair confusion, while idle time comes before typing error for low confusion. Though typing error is also the primary indicator of high and fair frustrations, other features are still needed before it is acknowledged as such. Lastly, there is a very slim chance to detect low frustration.

Keywords: Affect · Model · Novice programmer · Keyboard dynamics · Mouse behavior

1 Introduction

Affect is an observable expression of some emotional state [1–3]. It influences the ability of an individual to process information, to accurately understand and to absorb new knowledge [4].

In novice programmer studies, the negative affective states, particularly boredom and confusion, are negatively correlated with the student achievement while positive affect such as flow is positively correlated with achievement [5].

Affect detectors are built based on data acquired by sensors, human observations, or other peripherals. Several studies make use of keyboard dynamics as a source of affective data. These data include: typing speed, number of keystrokes, total time taken for typing, typing errors (the number of hits on the backspace key, delete key, or other unrelated keys), keyboard idleness [6, 7], keystroke latency time (dwell time) and keystroke duration time (flight time) between two-key (digraph or 2G) or three-key (trigraph or 3G)

© Springer International Publishing AG 2017
M. Numao et al. (Eds.): PRICAI 2016 Workshops, LNAI 10004, pp. 127–138, 2017.
DOI: 10.1007/978-3-319-60675-0_11

combinations [8, 9]. These studies examined further how these keystroke data are related to a generally described as positive and negative affective states.

A few studies also examined how mouse movements are related to irritation, annoyance, reflectiveness [10], and boredom [11]. Other studies also make use of the combined keyboard and mouse data to examine how these are related to affective states in terms of valence and arousal [7].

There are only a few studies that detect the affective states of novice programmers [e.g. 12, 13]. Also, there is no literature yet that uses the combined keyboard and mouse data to detect such states. This study hopes to contribute to the literature by building and validating a detector for negative affect of novice programming students using both the keyboard and the mouse data. We also attempt to answer the following research questions: (1) what are the notable features from keyboard dynamics and/or mouse behavior that help out in the recognition of negative affective states of novice programming students? (2) how is student's affect related to keyboard dynamics and/or mouse behavior; (3) are the notable features "stable" or "consistent" over student's programming time period? (4) how do these features differ or similar among high/medium/low incidences of boredom, confusion, and frustration? and (5) what is the effect of combining mouse behavior with the keystroke dynamic features in predicting student's affect compared when using keystroke features alone or mouse features alone?

This study hopes to contribute to the development of formal models of recognizing affective states of novice programmers, using the most common, low cost, non-intrusive computer devices such as the keyboard and the mouse. The discovered models or patterns to recognize negative affective states in this study may be used by computer scientists in developing computational systems that may automatically provide feedback to both teachers and students.

2   Related Works

Though there are different devices for affective states detection when using a computer, the keyboard and the mouse are the most commonly available, low-cost, and non-intrusive devices that could obtain affect indicators.

There were several studies that use only the keyboard as data source for affect detection. For example, Khanna et al. [6] extracted keystroke features: typing speed, four statistics (mode, standard deviation, variance and range) from the number of typed characters for a defined time interval, total time taken for typing, number of backspace hits and idle times from recorded key logs to detect positive, negative, and neutral state of a computer user. These keystroke data were gathered from participants who were asked to retype some fixed texts in different time in order to acquire keystroke information under different affect states. The corresponding affect is collected by asking the participants to describe and report their affective state while doing the task. The resulting dataset was then analyzed through some data mining algorithms such as SMO, MLP, and J48, They found out that the increase in the user typing speed relative to neutral state is an indicator of positive affect state while the decrease in the typing speed relative to neutral state is an indicator of negative affect.

An attempt to detect confusion and boredom states of novice programming students, Felipe et al. [12] extracted the same keystroke features used by Khanna et al. [6]. They also wanted to determine which of the extracted features could be indicators of the said affective states. The authors were permitted to collect video and key logs from students having programming activities. They reviewed every 20-second segment of the collected video logs and observe the student's behavior. They label affect by matching the corresponding observations from a checklist that describes affective states in terms of student's behavior. Results show that in a 20-second interval, keyboard inactivity in that time interval is the indicator of boredom state while confusion state was observed when the number of backspaces is greater than the idle time.

Tsui et al. [9] also used key duration time (key press to key release) and key latency time (from one key release event to the next key press) features to examine the difference between positive and negative affect states. The keystroke data were collected by asking each participant to type a fixed number sequence with a pen on the mouth. The affect is labeled based on the teeth condition (positive) and the lip condition (negative) of the participant while typing. They found out that the duration time significantly show the difference between the two opposite states.

The features used by Bixler and D'mello [16] to discriminate between natural occurrences of boredom, engagement, and neutral states are divided into four keystroke and timing features: relative timing (session and essay timings), keystroke verbosity (number of keys and backspaces), keystroke timing (latency measures) and pausing behaviors. These features were extracted from the key logs of participants who were asked to write an essay about some selected topics using a computer. Likewise, the affect was labeled by asking the participant to view every 15-second segment of his video log and has to make self-judgment on what affective state was present in him during each time segment. Results show that when the identified keystroke and timing features were combined with task appraisal and stable traits features, it yields to a higher accuracy rate in classifying emotions, specifically, between boredom and engagement.

There were also studies that explored mouse as data sources in affect detection. For example, Tsoulouhas et al. [11] extracted seven mouse movement features to detect emotional state, specifically boredom, of students who attend a lesson online. The said features are: total average movement speed, latest average movement speed, mouse inactivity occurrences, average duration of mouse inactivity, horizontal movements to total movements ratio, vertical movements to total movements ratio, diagonal movements to total movements' ratio, and the average movement speed per movement direction. They found out that the primary indicators of boredom are the average movement speed per movement direction and the mouse inactivity occurrences.

A more comprehensive study on affect detection in terms of its two dimensions was presented by Salmeron-Majadas et al. [7]. They evaluated the keyboard and mouse affective data to identify participant's affective states in terms of valence and arousal. They combined some previously presented keyboard indicators such as the keystroke indicators used by Khanna [6] and Bixler and D'Mello [16], and the digraph and trigraph used by Epp et al. [8]. Their mouse indicators were generated from the participant's mouse clicks, cursor movements and scroll movements. These include: the number of button presses (left, right and both), overall distance, distance the cursor has been moved (covered distance) between two button press events, between a button press and the following button release event, between two button release events and between a button release and the following button press events, the Euclidean distance in the previous described cases, the difference between the covered and the Euclidean distance between the events described before, and the time elapsed between the mentioned events. After the participants finished the given task, they were asked to evaluate and score their affective state using the SAM scale. They computed the correlation between the extracted mouse/keyboard indicators and the reported affective states and found out that the mouse indicators that are correlated to the valence dimension of affect are: the mean time between two consecutive mouse button press events; the mean time between two consecutive mouse button release events; the standard deviation of the difference between the covered and the Euclidean distance between two consecutive mouse button press events; the standard deviation of the difference between the covered and the Euclidean distance between a mouse button release and the following mouse button press events; and the mean time between a mouse button release and the following mouse press button event; while the keyboard indicators are: the standard deviation of the time between two key press events; the mean duration of the digraph; the mean duration between the first key up and the next key down of the digraph; the duration between two key press events when grouped in digraphs; and the mean time between two key press events. On the other hand, the mouse indicators that identify the arousal dimension of affect are: the mean of the difference between the covered and the Euclidean distance between a mouse button release and the following mouse button press events; the mean of the difference between the covered; and the Euclidean distance between two consecutive mouse button press events; while the keyboard indicators are: number of keys pressed; the numbers of alphabetical characters pressed; the mean of the duration of the second key of the digraphs; the duration of the third key of the trigraphs; and the standard deviation of the duration of the digraph. Finally, they used these mouse and/or keyboard indicators in training some classifiers in order for them to know the prediction rates in recognizing positive and negative valence dimension of the participants. Results show that for some well-known classifiers such as C4.5 and Naïve Bayes, keyboard indicators alone provided the higher prediction rates than the mouse data alone, and even the combination of the data sources. However, for some more complex classifiers such as Random Forest and AdaBoost, the combined mouse and keyboard indicators provided the highest prediction rates among all the results.

Though there are some few studies on the affective states of novice programmers (e.g. [3, 5, 12, 13]), to date, there is no literature yet that uses the combined keyboard and mouse data to detect some negative affective states of these novices.

3   Methodology

3.1   Participants

The participants in this study were 55 volunteers from first year students of a higher educational institution in Makati City. All of them were given waivers to parents or guardians, asking permission to let their child participate in the study. Hence, only those students with parent's/guardian's consent were allowed to participate.

At the time of the study, the students were enrolled in CS126 - Programming 1 with no or minimal background in C++. CS126 is a first year introduction to programming course using structured programming approach. Topics include: simple C++ syntax; program flow description; variables and data types; C++ operators; C++ control structures such as sequential, selection, and iterative structures; and functions.

3.2 Data Collection Methods and Instruments

With the consent of the school, we used a customized mouse-key logger, web cam, the MS Movie Maker, and the Dev-C++ Integrated Development Environment.

Before the student works on its programming activity, the web cam is already properly in place and turned-on. The mouse-key logger and the Movie Maker were set and running in the background and hidden from the student in order not to bother him/her while he/she is doing the programming activity.

The mouse-key logger captured the mouse motion, mouse clicks, and mouse scrolls and the key event logs while the web cam captured the facial expressions and body movements of the student (video logs). The Dev-C++ was used as the programming environment in doing the programming activities.

Data was collected from the participants where the problem is about selection constructs and loop constructs, respectively. Data recording took almost 3 h.

3.3 Data Processing

We mapped the mouse-key logs with the video logs in several steps: We first cleaned the data by removing segments in the mouse-key logs that had no corresponding video logs; then we extracted potential keystroke and mouse dynamic features identified in some previous works, plus other features that may influence affect detection, from the mouse-key logs. The result was a comma separated value (csv) file containing the keyboard and mouse dynamic features at every 15-second interval. This file was called the "incomplete dataset" since the affect labels were not yet attached. We also divided the video logs into 15-second video time segments that corresponded to mouse-key time segments in the incomplete dataset. Then, affect labeling on each video segment was done by three trained labelers, one was a graduate student serving as lead and the other two were college seniors with strong background in computer programming. They watched the video together and came to a consensus regarding the student's affective state based on the coding scheme in Table 1. If there were disagreements, they played the segment until they agreed. Video segments where the participants showed curiosities about being monitored through the camera or not seen in the video were marked "X". Finally, we mapped each label of the video segment in the incomplete dataset (Fig. 1), and the instances labeled with "X" were deleted.

Table 1. Affective state criteria

Affective states	Description
Boredom	Slouching and resting the chin in his/her palm; Yawning; Zoned out within the software; Looks uninterested/unfocused; Barely uses the mouse/keyboard; Slouching; Eyes wandering
Confusion	Scratching his/her head; Repeatedly looking at the same interface elements consulting with a classmate or a teacher; Flipping through lecture slides or note; Statements such as "Why didn't it work?"; Still engage with the software; Cannot grasp/experiencing difficulty with the material; On-task conversation; Pouts/Frowns/wrinkles brows/forehead; Nail biting; Lip biting; Lip slightly ajar
Frustration	Banging on the keyboard or pulling at his/her hair; cursing; statements such as "What's going on?!"; Scratching the back of his head; Rubbing his neck from behind; Scratching any part from his body; Changing his sitting position; Lips pulled inward; Raising the arms lifts sometimes up (or two arms- like throwing something in the air); Deep breath

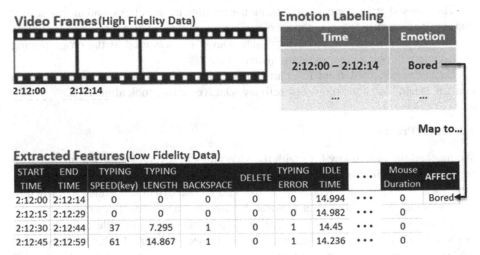

Fig. 1. Mapping of high fidelity data with the low fidelity data.

Determining of student's affect from the video segment was based on the modified coding scheme adopted from [3, 5, 14] and is presented in Table 1. The scheme was modified to find the state of confusion (negative valence, positive arousal), boredom (negative valence, negative arousal), frustrated, and a special affective state labeled as "others" [3, 6] in which the emotion with respect to the time frame was found to be neither confused, bored, nor frustrated.

The resulting complete dataset was then further divided into training and test set. Every fifth participant from the list was chosen as part of the test set while the rest were part of the training set.

3.4   Model Development and Data Analysis Methods

We used these datasets to develop several affective models for detecting confusion, frustration and boredom by training some well-known tree classifiers that could handle datasets with nominal class such J48, Decision Tree, and Random Forest using Rapid-Miner. Each classifier were trained and validated using different feature set, such as: keystroke verbosity features alone (KV); keystroke time duration and latency features of the digraph and trigraph alone (KT); all keystroke features - the combined verbosity, time duration and latency features (KF); mouse features alone (MF); and, the combined all keystroke and mouse features (KM). The gini index attribute criterion was used for feature selection and batch-X-validation to validate the model. The depth of the tree in each tree classifier was also explored in order to determine the model that has the highest performance in terms of accuracy rate and/or kappa statistic.

It was observed during the experiment that using keystroke time duration and latency features on the digraph and trigraph alone (KT), as well as mouse features alone (MF) do not provide a good model to detect negative affect since the kappa statistic is very low (less than 0.2) which implies a slight agreement [15]. It was also observed that the decision tree classifier consistently provide the highest kappa statistic and accuracy rate. It also implies that decision tree classifier gave the most acceptable model.

Lastly, the kappa and accuracy of the other feature sets are statistically tied (Table 2, columns 2 and 3). And since the kappa is in moderate agreement [15], it implies that these feature sets can be used to model negative affect detector. The models generated by the decision tree classifier for the said feature sets were tested using a pre-labeled test set for further investigation. The result of the tests is also presented in Table 2, columns 4 and 5. The table shows that the kappa and the accuracy significantly increased but are still statistically tied. This confirms that the three (3) feature sets can be used to model negative affect detectors of novice programming students.

Table 2. Model performance using decision tree classifier

Feature set	Training phase		Testing phase	
	Kappa	Accuracy (%)	Kappa	Accuracy (%)
KV	0.493	70.80	0.564	74.08
KF	0.489	71.03	0.568	74.28
KM	0.490	71.06	0.567	74.23

The tree models were further analyzed to find the significant features that help out in the recognition of negative affective states of novice programming students and how these features are related to student's affect. This was done by listing the unique inner nodes of the decision tree models generated by the classifier.

Using correlations in RapidMiner, it was observed that some of the notable features in the tree are strongly correlated. For example: typing error is highly correlated with backspace; total keyevents is also highly correlated with typing speeds and total time for typing; the sum of all time durations the student acted on the 1st key of the digraph (SUM_2G_1Dur) is fairly correlated with the maximum value in the set of the durations of the 1st key of the trigraph (MAX_3G_1Dur); and the total distance travelled by the mouse along the x-axis (MM_Total_X) is highly correlated with mouse activity duration. Thus, to achieve a more parsimonious model, we tried iteratively removing some features that are highly correlated to other features. Results show that the kappa and accuracy slightly improved (see Table 3). The table shows that kappa and accuracy in all the feature sets are almost equal. It implies that the notable features from the keystroke verbosity feature set alone (KV) or the combined verbosity, duration, and latency keystroke features (KF) are already enough to model a negative affect detector of novice C++ programming students. However, adding MM_Total_X (total distance travelled by the mouse along the x-axis) mouse feature with the keyboard features (KF) slightly improved the recognition rate of the model (Table 3).

Table 3. Model performance when some features correlated to other features were removed.

Feature set	Kappa		Accuracy (%)		Remaining significant features
	Before removal	After removal	Before removal	After removal	
KV	0.564	0.569	74.08	74.23	Typing error, typing variance, idle time, total key events, F9
KF	0.568	0.568	74.28	74.28	Typing error, typing variance, idle time, total key events, F9, MAX_3G_1Dur, AVE_3G_2D3D, and SUM_2G_1Dur
KM	0.567	**0.572**	74.23	**74.37**	Typing error, typing variance, idle time, total key events, F9, SUM_2G_1Dur, AVE_3G_2D3D, and MM_Total_X

To specifically determine how student's affect related to keyboard and mouse dynamics, the unique paths from the root of the decision tree of the KM feature set, to the its leaves were analyzed and then transformed into rules. The result is shown in Table 4.

We examined if the features were stable over time since it is possible that student keyboard and mouse dynamics change as the student develops and completes a program. Also, a student may type more in the beginning of the development process, when he is still writing code, and less so when he is debugging. We therefore divided the dataset into the first 1/3, the second 1/3, and the last 1/3 of the observation period and reprocessed each subsets.

Results show that when students are just starting with their programming activity (first 1/3 of the period), the most notable features that determines student's negative affect in all feature sets are the typing error, and idle time. Though typing error and the idle time are also the dominant features on the second 1/3 of the period, other keystroke verbosity features such as typing variance, total keyevents, and the number of times the student presses F9 (shortcut to compile and run the program) are included. Also, adding the average duration time of the first key in the trigraph (AVE_3G_1Dur) or the total distance travelled by the mouse along the x-axis (MM_Total_X) improves the recognition rate. Lastly, at the time the programming period is almost toward its end (last 1/3 of the period), the typing error and idle time are still the dominant features, but adding the typing variance and total mouse movement along the x-axis increases student's negative affect detection. It was also observed that the total keyevents and the average duration time of the first key in the trigraph (AVE_3G_1Dur) that represent the

Table 4. How student affect related to keystroke dynamics and mouse behaviors.

Affect (most likely)	Pattern (based on a 15-second observation)
Boredom	IF ((Typine error ≤ 0.50) and (Idle time > 14.98) and (Total keyevents ≤ 0.50) and (MM_Total_X ≤ 243.50))
Frustration	IF ((Typing error ≤ 0.50) and (Idle time > 14.98) and (Total Keyevents > 4.50))

Confusion	IF (Typing error > 3.50); OR
	IF ((1.50 < Typing error ≤ 3.50) and (Typing variance > 0.815) (MM_Total_X > 17383.50)); OR
	IF ((Typing variance > 0.815) and (Typing error ≤ 1.50) and (SUM_2G_1Dur ≤ 0.83)); OR
	IF((1.5 ≤ Typing error ≤ 3.50) and (Typing variance ≤ 0.815)); OR
	IF ((Typing variance ≤ 0.815) and (Typing error ≤ 1.50) and (AVE_3G_2D3D ≤ 5.170)); OR
	IF ((Typing error ≤ 0.50) and (Idle time > 14.98) and (Total Keyevents ≤ 0.50) and (MM_Total_X > 850)); OR
	IF ((Typing error ≤ 0.50) and (Idle time ≤ 14.98) and (Total Keyevents > 9.50) and (SUM_2G_1Dur > 0.924) and (F9 > 27)); OR
	IF ((Typing error ≤ 0.50) and (Idle time ≤ 13.636) and (Total Keyevents ≤ 9.50) and (MM_Total_X > 872956.5))

Table 5. Differences or similarities among high/medium/low incidences of boredom, confusion, and frustration.

Affect		Keystroke dynamic features	Combined mouse and keystroke dynamics features
Boredom	High	Idle time, Total keyevents	Idle time, mouse activity duration
	Fair	Idle time, Total keyevents	Idle time, total keyevents
	Low	Typing error, Idle time	Typing error, idle time, numbers, mouse activity duration
Confusion	High	Typing error	Typing error
	Fair	Typing error, Typing speed(char)	Typing error, typing speed(char)
	Low	Idle time, Typing error	Idle time, typing error
Frustration	High	Typing error, Typing variance, Idle time, Control	Typing error, typing variance, idle time, control
	Fair	Typing error, Idle time, Keypress_DeltaX, F9, etc.	Typing error, idle time, CMM_time, keypress, etc.
	Low	(no sign of low frustration)	(the tree is very deep where there is a very minimal sign of frustration)

movements of the keys, including F9 which represents running the program were gone towards the end of the programming period. This may indicate that there were only few monitored keyboard activities. Probably, some of the students may have stopped working; either they are already finished with the activity or they have abandoned their work.

Finally, to determine how the notable features differ or similar among high/medium/low incidences of boredom, confusion, and frustration, the original dataset was divided into other subsets by computing the percentage of the time each student was observed to be bored, confused or frustrated and then segregate the data into the top 1/3 of those who are bored, confused or frustrated, the middle 1/3, and the lowest 1/3, and then re-process each subsets. The result is shown in Table 5.

4   Conclusion

This study was conducted to address the following research questions: (1) what are the notable features from keyboard dynamics and/or mouse behaviors that help out in the recognition of negative affective states of novice programming students? (2) how is student's affect related to keyboard dynamics and/or mouse behaviors; (3) are the notable features "stable" or "consistent" over student's programming time period? (4) how do these features differ or similar among high/medium/low incidences of boredom, confusion, and frustration? and (5) what is the effect of combining mouse features with the keystroke features in predicting student's affect compared when using keystroke dynamic features alone or mouse behaviors alone? These questions are answered as follows:

(1)   The notable features from keyboard dynamics and/or mouse behaviors that help out in the recognition of negative affective states of novice programming students are presented in Table 3. These include: the student's typing errors incurred (the number times the backspace and delete keys were pressed); the length of time the student is idle (not pressing any key in the keyboard); the student's typing variance (his/her typing varies with time); the number of key events (keydown + keypress + keyup) he/she executed in the keyboard; total distance the student moved the mouse along the x-axis (MM_Total_X); the sum of all time durations the student acted on the 1st key of the digraph (SUM_2G_1Dur); the average time duration between the 2nd and 3rd keydown of the trigraph (AVE_3G_2D3D); and, the number of times F9 key (shortcut to compile and run the program) was pressed.

(2)   As shown in Table 4, student's boredom is related to both keystroke dynamics and mouse behavior. The keyboard has almost no activity while the mouse has a very minimal movement along the x-axis. On the other hand, student's frustration is similar to boredom, except for the mouse features, since for this affect, students tend to release the mouse and scratch their head or do some other hand gestures. There is almost no keyboard activity too since when a student get frustrated, he/she usually pause for a while and do nothing. Lastly, student's confusion is both related

to keystroke dynamics and mouse behavior. The table shows that there are several indicators when a student is confused.

(3) After analyzing the data at first 1/3, the second 1/3, and the last 1/3 of the observation period, it was observed that the features are not stable since there are many features needed in detecting negative affect at the middle (second 1/3) of the observation period. It was also observed that the typing error has the greatest influence during the first 1/3 and second 1/3 of the observation period followed by the idle time, while the idle time has the greatest influence during the last 1/3 followed by the typing error.

(4) Table 5 shows that idle time has the greatest influence in detecting high and fair boredom but it is just secondary with the typing error for low boredom. On the other hand, typing error has the greatest influence in detecting high and fair confusion but it is just secondary with the idle time for low confusion. Though typing error is also the primary indicator of high and fair frustrations, it requires other features before it is acknowledged as such.

As shown in the last row of Table 3, adding a mouse feature, particularly with the distance it travelled along the x-axis, with the keystroke features improve the detection of student's affect compared when using keystroke dynamic features alone or mouse behaviors alone.

References

1. Affect. Encyclopedia of Mental Disorders. http://www.minddisorders.com/A-Br/Affect.html
2. Psychiatry Clerkship. http://depts.washington.edu/psyclerk/glossary.html
3. Carlos, C.M., Delos Santos, J.E., Fournier, G., Vea, L.: Towards the development of an intelligent agent for novice programmers through face expression recognition. In: Proceedings of the 13th Philippine Computing Science Congress, pp. 101–106 (2013)
4. Picard R.W., The Medial Lab – Affective Computing Group: Affective computing. http://affect.media.mit.edu/
5. Rodrigo, M.M.T., Baker, R.S., Jadud, M.C., Amarra, A.C.M., Dy, T., Espejo-Lahoz, M.B.V., Lim, S.A., Pascua, S.A., Sugay, J.O., Tabanao, E.S.: Affective and behavioral predictors of novice programmer achievement. In: Proceedings of the 14th Annual ACM SIGCSE Conference on Innovation and Technology in Computer Science Education (ITiCSE 2009), vol. 41, no. 3, pp. 156–160 (2009). http://doi.acm.org/10.1145/1562877.1562929
6. Khanna, P., Sasikumar, M.: Recognising emotions from keyboard stroke pattern. Int. J. Comput. Appl. **11**(9), 1–5 (2010)
7. Salmeron-Majadas, S., Santos, O.C., Boticario, J.G.: Exploring indicators from keyboard and mouse interactions to predict the user affective state. In: Educational Data Mining (2014)
8. Epp, C., Lippold, M., Mandryk, R.L.: Identifying emotional states using keystroke dynamics. In: Proceedings of the SIGCHI Conference on Human Factors in Computing Systems, pp. 715–724 (2011)
9. Tsui, W.H., Lee, P., Hsiao, T.C.: The effect of emotion on keystroke: an experimental study using facial feedback hypothesis. In: 2013 35th Annual International Conference of the IEEE Engineering in Medicine and Biology Society (EMBC), pp. 2870–2873 (2013)

10. Schuller, B., Rigoll, G., Lang, M.: Emotion recognition in the manual interaction with graphical user interfaces. In: Proceedings of the 2004 IEEE International Conference on Multimedia and Expo (ICME 2004), vol. 2, pp. 1215–1218 (2004). http://dx.doi.org/10.1109/ICME.2004.1394440

11. Tsoulouhas, G., Georgiou, D., Karakos, A.: Detection of learner's affective state based on mouse movements. J. Comput. **3**(11), 9–18 (2011)

12. Felipe, D.A.M., Gutierrez, K.I.N., Quiros, E.C.M., Vea, L.A.: Towards the development of intelligent agent for novice C/C++ programmers through affective analysis of event logs. Proc. Int. MultiConf. Eng. Comput. Sci. **1**, 511–518 (2012)

13. Lee, D.: Detecting confusion among novice programmers using BlueJ compile logs. Master's thesis. Ateneo de Manila University, Quezon City (2011)

14. Dragon, T., Arroyo, I., Woolf, B.P., Burleson, W., el Kaliouby, R., Eydgahi, H.: Viewing student affect and learning through classroom observation and physical sensors. In: Woolf, B.P., Aïmeur, E., Nkambou, R., Lajoie, S. (eds.) ITS 2008. LNCS, vol. 5091, pp. 29–39. Springer, Heidelberg (2008). doi:10.1007/978-3-540-69132-7_8

15. Viera, A.J., Garrett, J.M.: Understanding interobserver agreement: the kappa statistic. Fam. Med. **37**(5), 360–363 (2005)

16. Bixler, R., D'Mello, S.: Detecting boredom and engagement during writing with keystroke analysis, task appraisals, and stable traits. In: Proceedings of the International Conference on Intelligent User Interfaces, pp. 225–234 (2013)

Affective Laughter Expressions from Body Movements

Jocelynn Cu[(✉)], Ma. Beatrice Luz, McAnjelo Nocum, Timothy Jasper Purganan, and Wing San Wong

Center for Human Computing Innovations, De La Salle University, Manila, Philippines
{jocelynn.cu,ma_beatrice_luz,mcanjelo_nocum,
timothy_purganan,wing_wong}@dlsu.edu.ph

Abstract. The main goal of this study is to classify affective laughter expressions from body movements. Using a non-intrusive Kinect sensor, body movement data from laughing participants were collected, annotated and segmented. A set of features that include the head, torso, shoulder movements, as well as the positions of the right and left hands, were used by a decision tree classifier to determine the type of emotions expressed in the laughter. The decision tree classifier performed with an accuracy of 71.02% using a minimum set of body movement features.

Keywords: Affective laughter · Laughter expression · Analysis of body movement · Gestures

1 Introduction

Laughter is an innate human emotional expression more commonly associated with positive feelings. People often laugh when they feel happy, or excited. But they also laugh when they feel embarrassed, or sad. Recent studies have shown that there is a wider range of emotions that are expressed through laughter. In particular, the study of [5] has identified 25 laughter types associated with both positive and negative emotions, and 6 laughter types not necessarily used to express emotions.

The work of [2] classified affective laughter into five types, i.e., happiness, excitement, giddiness, embarrassment, and hurtful, by analysing facial and vocal cues. Their study, however, did not include body movement and gestures in characterizing different types of laugher.

Griffin and his colleagues [3] identified perceptible body movements that can be used to distinguish different types of natural laughter. Then from these body movements, they applied supervised machine learning techniques to automatically classify different types of laughter. Motion data were recorded from 9 participants (3 males and 6 females) wearing motion capture suits in a standing and sitting positions. Samples were taken when the participants are performing specific tasks, such as playing word games, and during conversations between tasks. Participants were also requested to produce fake laughter. Samples are initially labelled as laughter or non-laughter, with laughter types further classified as hilarious, social, awkward, and fake. Based on the body movement analyses of 32 observers, the study was able to determine that the hands, shoulders, spine and neck movements are useful discriminating features. Supervised learning models,

© Springer International Publishing AG 2017
M. Numao et al. (Eds.): PRICAI 2016 Workshops, LNAI 10004, pp. 139–145, 2017.
DOI: 10.1007/978-3-319-60675-0_12

which include k-Nearest Neighbor (kNN), Multi Layer Perceptron (MLP), Random Forest (RF), Linear and Kernel Ridge Regression (RR, KRR), Linear and Kernel Support Vector Regression (SVR, KSVR), were built to automatically classify laughter types. Recognition results show that the RF model outperforms the other models in terms of accuracy in classification. The model was also able to distinctly classify hilarious laughter, social laughter, and non-laughter.

The work of Niewiadomski and his colleagues on [6], on the other hand, is focused on showing the robustness of the body as a cue in discriminating laughter expressions from non-laughter expressions. This study developed a real-time system prototype to demonstrate this potential. For the first part, data from ten participants (8 males and 2 females) wearing motion capture suits were collected. Participants are allowed to converse in their native language while performing specific tasks, such as watching comedies or playing games. Two raters were engaged to annotate the data in the absence of audio information. A feature vector that captures full body movement, consisting of 13 features (F1–F13) describing head movements (left, right, front, back), weight shifting, knee bending, abdomen, trunk, arm, and shoulder movements, was computed from the motion capture data. Five learning models were built and tested to evaluate how well the discriminative algorithms (SVM, kNN, RF) and probabilistic algorithms (Naïve Bayes (NB), Logistic Regression (LR)) work on the data. Based on experiment results, it was confirmed that discriminative classifiers outperforms probabilistic classifiers. The second part of this study focused on building a real-time system prototype, in which a Kinect sensor was used to body movement, 9 features (K1–K9) were computed. These features track the head, torso, and shoulder movements. Features that tracks legs and arms movement were not computed since the participants in this set-up are in a sitting position. SVM was used classifier for the prototype.

Similar to [6], this study also built a real-time system prototype that captures body movement for laughter type classification. We also used Kinect sensor to capture motion data. However, in addition to the head, torso, and shoulders, we included the hand position in reference to the head position, as part of our feature set. With this, we attempted to classify more than two types of affective laughter. This paper describes the methodology we employed and the results of our work.

This paper is organized as follows: Sect. 2 describes the methodology we used to build the real-time system prototype, which includes the creation and annotation of our training and test data, the computation of our feature set, and the modelling tasks; Sect. 3 presents the results of our experiments and the discussion of these results; and, our conclusion in Sect. 4.

2   Methodology

In this study, our goal is to investigate the possibility of using an accessible device like the Kinect sensor to capture full body movement, and with these less precise features, try to differentiate affective laughter expressions into one of the following types [7]: happiness, giddiness, excitement, embarrassment, and hurtful.

2.1   Data Collection and Annotation

Data was collected using Microsoft Kinect 360, to capture full body movement, and a Sony HD video recorder, to capture the entire session for annotation purposes. Groups of 2 to 3 friends, aged 17 to 20 years old, a mix of males and females, were invited to the data collection sessions. Recording sessions last from 5 min to 20 min, depending on how comfortable the participants are talking about a particular topic. To make sure that full body movement is captured, the target participant is asked to stand 1.5 m away facing the other person, where the Kinect sensor is also positioned. The only instruction given to the participants is that they choose and talk freely about a topic from a given list. They were not given any time limit. The actual list of topics given to the participants include their personal interests or hobbies, embarrassing moments in high school or college, inside jokes of their group, secret crush or love life. Among these topics, talking about their crush and love life generated a lot of laughter segments. Incidentally, so are telling obscene jokes, insulting each other, or insulting a common friend.

Manual segmentation resulted to 245 laughter segments (around 72 min of recording) collected from 9 participants (6 males, 3 females). Non-laughter segments, speech-laughs, and fake laughs were removed. Four raters who got an above average score (>49) in the Baron-Cohen Empathic Quotient (EQ) Test were asked to review and annotate the laughter segments. These raters are not acquainted with the participants and they have no idea what the participants are talking about. The annotation process requires the raters to watch the video, without sound, and to identify which of the five types of affective laughter is evident in the clips. If the laughter segment cannot fit into one of the five categories, it will be marked for removal. To obtain a unified result, the inter-rater agreement was computed using Fleiss' Kappa, which is 0.57 and is considered "moderate agreement". The process resulted to 217 laughter segments distributed as follows: 71.4% happiness, 23% embarrassment, 3.2% hurtful, 1.8% giddiness, and 0.5% excitement. Table 1 shows the distribution.

Table 1.   Laughter segment distribution based on types of affective laughter.

Laughter type	# of segments
Happiness	155
Embarrassment	50
Hurtful	7
Giddiness	4
Excitement	1
TOTAL	217

Brief interview with the raters revealed that aside from the head, torso, and shoulder, they also rely on the hand position to help them determine the laughter types. Since the participants are not moving around while laughing, points on the knee and the foot were ignored and not included in the analysis.

2.2 Feature Extraction

The Kinect sensor collects 13 points from the body, as shown in Fig. 1. These points are located at the head, left shoulder, center shoulder, right shoulder, left elbow, left hand, spine, right elbow, eight hand, left knee, right knee, left foot, and right foot.

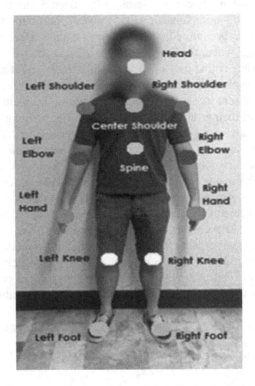

Fig. 1. Placement of body points as determine by the Kinect sensor.

Based on the recommendations of the raters, we computed the features from movements of the head, torso, shoulders and position of the hands. The formulas were taken

Table 2. Summary of body movement features used in this study in comparison with that of Niewiadomski et al. in [6].

Body part	Movement/Position	Niewiadomski et al. [6]
Head	Head tilt (left or right)	F1 – Head side movement
	Head nod (downward or upward)	F2 – Head front and back movement
Torso	Moving (forward or backward)	F6 – Trunk straightening movement
Shoulders	Leaning (forward or backward)	F7 – Trunk rocking movement
	Shrugging (shrug or no shrug)	F13 – Shoulder shaking movement
Hand	Left and Right Hand Positions (above head, on head/face, below head)	None

from the works of [1] on real-time academic affective states recognition from body movements and gestures, and of [4] on control methods in a 3D space using Kinect. Table 2 compares the features used in this study and the features used by Niewiadomski et al. in [6].

2.3 Classification

Three classification algorithms were tested on the data prior to implementation on the prototype. These are the k-Nearest Neighbor (kNN), Decision Trees (J48), and Naïve Bayes (NB). Weka was used to build the models for comparison and to do the 10-fold cross validation on the results.

The Bi-Directional Best Fit First Feature Selection algorithm was applied on the feature set to determine which features are useful and should be included in the proto-type.

3 Results and Discussion

Due to the limited number of segments for the other laughter types, the automatic classification task was reduced to binary classification between happiness and embarrassment laughter. A dataset with equal number of happiness and embarrassment segments was built using the complete feature set. The result is shown in Table 3.

Table 3. Summary of classification tests using 10-fold cross validation.

Algorithm	Complete feature set			Reduced feature set		
	Accuracy	Kappa	RMSE	Accuracy	Kappa	RMSE
NB	0.5284	0.4105	0.3485	0.5272	0.4090	0.3486
kNN (n = 5)	0.6628	0.5785	0.3092	0.6670	0.5838	0.3090
J48	0.7156	0.6445	0.2978	0.7102	0.6377	0.2997

The complete feature set includes the head tilt, head nod, torso movement, leaning, shrugging, right hand position, and left hand position. The reduced feature set was generated by applying the feature selection algorithm on the complete set to determine minimal set of features useful in discriminating laughter types. This resulted to a set of features that considers only the head tilt, torso movement, and the positions of the right and left hands. In terms of accuracy, the reduced feature set did not perform better than the complete feature set. However, in a system that is expected to do a classification in real-time, a reduced feature set is attractive because it implies reduced computational load.

Intuitively, the features that were retained for classification were also evident in the general observations of the raters. Head tilting movements (i.e., either to the left, or to the right) combined with hands on the head (i.e., either covering one's face, or touching one's hair) can differentiate when one is happy or embarrassed, as shown in Fig. 2.

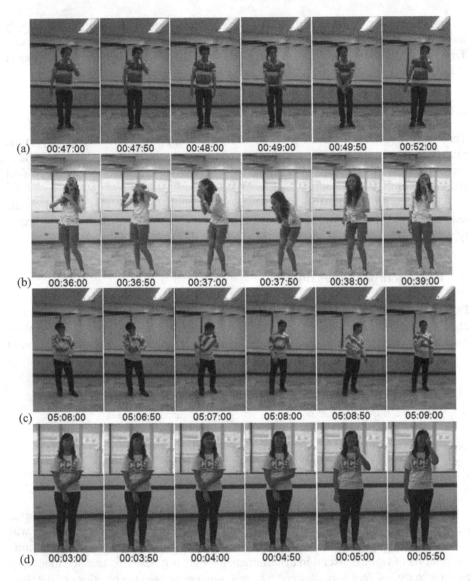

Fig. 2. (a) and (b) show male and female participants expressing happy laughter type. (c) and (d) show male and female participants expressing embarrassed laughter type.

Majority of the participants shrug their shoulders while laughing. However, in the data, it seems that the movement is minute which makes it undetectable by the machine. Head nodding is another movement that is uncommon for either genders. Leaning the body forward while laughing is observed in only 1 female participant. This is also the same participant who produced the most number of laugh segments.

Regardless of the number of features used in the classification task, kNN and J48 performed better than NB, which is consistent with the findings of [3, 6].

4 Conclusion

We set out to investigate the possibility of using a non-intrusive but less precise sensor like the Kinect to capture full body movement, and use these to differentiate affective laughter expressions into five classes.

Unfortunately, we were unable to collect enough laughter segments that express hurt, giddiness and excitement to appropriately explore this problem. This lack of data reduces the problem into a binary classification task. However, the classification results support the work of [6] in that it is indeed possible to use body movements as cue to differentiate laughter types, in the absence of other modalities like facial expressions or vocalizations.

Moreover, of the 217 laughter segments, 48.57% were produced by the three female participants, while the remaining 51.43% were produced by the six male participants. This may have also affected the laughter models of the classifiers.

The Kinect sensor was able to collect enough points on the body to extract useful body movement features. Based on our feature selection task, it seems useful to also include the position of the left and right hands in differentiating laughter types.

With a wide range of emotions associated with laughter, a multidimensional description of affective laughter, in conjunction with contextual information at which the laughter was expressed, will complicate the classification task but improve the accuracy of laughter type classification.

References

1. Cheung, O.H.: Real time academic emotion recognition using body gestures. Masters thesis, De La Salle University (2012)
2. Galvan, C., Manangan, D., Sanchez, M., Wong, J., Cu, J.: Audiovisual affect recognition in spontaneous Filipino laughter. In: 2011 Third International Conference on Knowledge and Systems Engineering (KSE), pp. 266–271. IEEE (2011)
3. Griffin, H.J., Aung, M.S.H., Romera-Paredes, B., McLoughlin, C., McKeown, G., Curran, W., Bianchi-Berthouze, N.: Laughter type recognition from whole body motion. In: 2013 Humaine Association Conference on Affective Computing and Intelligent Interaction, pp. 349–355. IEEE (2013)
4. Kang, J.W., Seo, D.J., Jung, D.S.: A study on the control method of 3-dimensional space application using Kinect system. Int. J. Comput. Sci. Netw. Secur. 11(9), 55–59 (2011)
5. McKeown, G., Cowie, R., Curran, W., Ruch, W., Douglas-Cowie, E.: ILHAIRE laughter database. In: Workshop on Emotion Sentiment and Social Signals LREC, Istanbul, pp. 32–35 (2012)
6. Niewiadomski, R., Mancini, M., Varni, G., Volpe, G., Camurri, A.: Automated laughter detection from full-body movements. IEEE Trans Hum. Mach. Syst. 46, 113–123 (2015)
7. Suarez, M.T., Cu, J., Sta, M.: Building a multimodal laughter database. In: LREC 2012, Istanbul (2012)

RSAI 2016: Workshop on Research Student Symposium on Artificial Intelligence and Applications

Real-Time Snoring Sound Detecting U Shape Pillow System Using Data Analysis Algorithm

Patiyuth Pramkeaw$^{(\boxtimes)}$, Penpichaya Lertritchai,
and Nipaporn Klangsakulpoontawee

Department of Media Technology,
King Mongkut's University of Technology, Thonburi, Thailand
patiyuth.pra@kmutt.ac.th,
bewwylert@gmail.com, mmayo643@gmail.com

Abstract. This paper aims to design and build snoring sound detecting u-shape pillow. Research operating includes four steps as (1) to study the problem of snoring, (2) to analyze the related information for develop the snoring sound detecting u shape pillow with designing the structure and sensory circuit inside the pillow. The u-shape has been designed as a neck supporter for user. The main part of project is the module having microphones that receive a sound of snoring. When a snoring sound was detected, the module will command the vibrating motor to work and alert the user to change his/her body posture. This change will help user stopping a snoring, which controlling by the C language programs, (3) to assess the quality of pillow detect snoring by five experts, (4) as result shown, the proposed pillow can detect snoring sound at 80% of accuracy based on testing with three different users.

Keywords: Snoring sound · Neck pillow · Detect snoring sound

1 Introduction

Snoring is a problem that is commonly found in people between 30–35 years old, often occurs to males rather than females, and basically increases with ages. It can be simple snoring which has no harms but social effects as well as impacts on others' quality of life, especially bedfellows', owing to its disturbing sounds. The other type of snoring that can happen is snoring with obstructive sleep apnea. Patients of this type normally face dogsleep and waking up frightened intermittently, resulting in poor sleep. These patients, therefore, tend to work inefficiently and may have traffic accidents on account of drowsy driving [1]. Or even those who have this state while working with machines in factories can also have a high risk of dangers.

However, most people usually overlook snoring as they think it is a normal symptom which causes no severe perils. In fact, if they let snoring continue without treatment [2], it will lead to unpleasant situations with their bedfellows or people around them. To make matters worse, it might bring about jeopardies to the daily life, work performance, health, and risks of any other diseases to death [3].

According to the background and the significance of the mentioned problems, the research were interested in the technology to solve these problems. Relevant principles

© Springer International Publishing AG 2017
M. Numao et al. (Eds.): PRICAI 2016 Workshops, LNAI 10004, pp. 149–159, 2017.
DOI: 10.1007/978-3-319-60675-0_13

as well as theories were explored in order to create a "sensor pillow" or a "snoring sensor pillow." The main objective was to relieve a patient with snoring symptom by applying the microphone module to detect his/her snoring sound. After the sound is detected, the vibrating motor will vibrate the pillow so as to motivate the consciousness of the sleeper and stop his/her snoring. The pillow is easy to use without electricity that regularly possesses high risks and hazards. It is also small and portable, and thus suitable for a patient with snoring problem and has to travel a lot.

2   Snoring

Snoring is a state or a symptom with abnormal volumes of breathing sounds during sleep. It is one of the symptoms that normally happen to people at all ages from children to the elderly. The older they are, the more symptom increases. Several studies revealed that the symptom was found in males at 24–30% whereas approximately 15% in females. And when they reach the ages of 60–65, the symptom raises up to 60% in males and 40% in females [2].

2.1   Primary Snoring

Primary Snoring do not badly affect health. It merely causes annoyance to people nearby and partial airway obstruction. That is because while you are sleeping, it is the time for muscle relaxation, including pharynx. Your tongue and uvula fall behind, particularly when you lie on your back [2]. The airway or intrarespiratory at this part becomes narrower. So, when you breathe in through this narrow position, your uvula and soft palate or root of tongue are vibrated. This is how snoring occurs (Fig. 1).

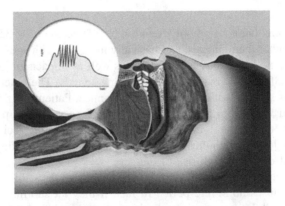

Fig. 1. Shows primary snoring

2.2   Obstructive Sleep Apnea: OSA

The symptom takes place due to the extremely narrow airway, possibly because of very narrow pharynx. For example, there is soft palate tissue, uvula, or huge flabby root of tongue; or enlarged tonsil gland that obstructs pharynx. Some people have little face bones or molars, so the back of their airways are narrower than usual, including those who have shorter chins, of which tongues fall behind deeper then normal people. These group of patient basically produce inconsistent snoring sounds. To clarify, they project both loud and soft snoring sounds intermittently, and they will keep snoring more loudly. Then, they will stop snoring for a moment or it is call "obstructive sleep apnea." It is regarded as a hazardous moment because oxygen levels drop down, acting upon malfunctions of some of their organs such as lungs, hearts, and brains. Their bodies will respond to this state automatically afterwards, that is, their brains are stimulated and thus subjected patients are awakened from sleep to regain their breathing [2]. They usually wake up frightened like being taken aback or chocking on their own saliva. Or they might breathe hard to get oxygen as if to recover from suffocation. Soon after, their brains will fall asleep again, and their breathing is obstructed again. Then, the brains need to be aroused once more, and their sleep is impeded again, too. Such incident keeps circulating repeatedly. As a result, the sound sleep of patients with inconsistent snoring is relatively deficient [5]. They, therefore, always wake up with the feeling of insufficient sleep despite fine numbers of sleeping hours, and it also leads to damages of health, especially their hearts, blood circulation system, lungs, and brains (Fig. 2).

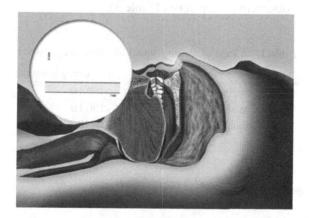

Fig. 2. Obstructive sleep apnea

2.3   A Kind of Snoring Sound

When concentrating on recording with tools or recording equipment from various reports, the researcher found that snoring sounds can be divided into some major characteristics, i.e., sound properties (e.g., loudness and frequencies); simple snoring sounds and obstructive sleep apnea sounds; and snoring sounds from different sources — from soft palate and root of tongue. The collected information helps noticing 8 characteristics of snoring sounds as per below:

(1) Frequencies of snoring sounds are between 2 up to 50 Hz, and the loudness can be 20 dB up to 85 dB.
(2) Snoring sounds of patients with simple snoring are with fundamental frequencies and harmonic pattern, and their snoring sounds are higher than 150 Hz with wide bandwidths.
(3) Snoring sounds of patients with obstructive sleep apnea are not harmonic pattern. The sounds are with low frequencies (<150 Hz) with narrow bandwidths.
(4) Snoring sounds from soft palate are in the well-organized wave form. Waves are repeated every 10–30 ms. Their peak frequencies are low (285 Hz on average), and often found more than from root of tongue.
(5) Snoring sounds from root of tongue initiate disorganized wave form, with high peak frequencies (885 Hz on average).
(6) Snoring sounds from soft palate are like flapping noises whereas those from root of tongue are like stridor.
(7) Flexible nasopharyngoscopes can be manipulated to investigate any specific spots of the vibrating upper airway that causes snoring sounds while patients are sleeping or under sleeping pills. This is why the diagnosis is called "sleep nasendoscopy."
(8) Very loud snoring sounds and constant pauses relate to sleep apnea. And when patients breathe again, the sounds emerge again as well. This is mostly found in patients with obstructive sleep apnea (Table 1).

Table 1. A kind of snoring sound and frequency (Hertz)

A kind of snoring sound	Frequency (Hertz)
Primary snoring	>150 Hz
Obstructive sleep apnea	<150 Hz
Snoring sounds from soft palate	285 Hz
Snoring sounds from root of tongue	885 Hz

3 Literature Review

3.1 Sensor Pillow System [12]

"Sensor pillow system" embraces the development of diagnosis system of sleep disorders among paralyzed patients. The system basically examines hearts, respiratory system, and reactions during sleep. They are all up to polysomnography and intimate care from physicians. Zigbee or GSM polysomnography will record brain waves, oxygen levels in blood, and pulse rates from body movements. They can be measured by using sensor pillows which comprise of the management of FSR sensors, Zigbee or GSM which emits analog signal. The entire management of FSR sensors, respiratory system sensors, Zigbee, or GSM with Visual Basic (VB) software can be utilized to check/diagnose sleep disorders, pulse rates, and blood pressures.

3.2 Short-Term Outcomes of Transoral Radiofrequency Somnoplasty Treatment for Snoring [9]

For the results of short-term treatment of snoring symptom through oral radiofrequency (RF), it will be elaborated next. This research gathered the data/information of patients with snoring who were treated by oral radiofrequency from Department of Otolaryngology, Maharaj Nakorn Chiang Mai Hospital. The results of polysomnography were also embraced. After that, the data was analyzed and compare between pre-treatment and post-treatment (after 3 months of treatment). The results indicated that there were 34 patients, 24 of them were males. Mean of the violence of snoring sounds before treatment was 7.0, and down to 3.6 after treatment (reduced by 48.7%). Mean of daytime drowsiness before treatment was 8.4, and dropped down to 5.4 after treatment. To conclude, the application of oral radiofrequency could efficiently diminish the violence of snoring sounds as well as daytime drowsiness. It also generated less pain to wounds, with only minor mild complications. Thus, this can be one of the great alternatives to cure patients with this problem.

4 Methodology

4.1 Sensor and System

Figure 3, presented the sound sensor module as an electret microphone amplifier with frequencies between 20 Hz to 20 kHz. The module also had the extended circuit with IC MAX4466, which was a low-noise microphone amplifier that could cut off all noises [6]. A tiny trimmer pot was included and could adjust the gain from $25\times$ to $125\times$ which produced the value around 200 MVpp, and possessed an intrinsic ability to cut noises. It was used to detect snoring sounds, with rail-to-rail outputs, and was connected to a 5-volt DC power supply [7]. The noise was projected out of output's legs with DC Bias at VCC2. The motor was vibrated to stimulate the consciousness of snoring patients so that they would stop snoring as shown in Fig. 4, and the frequency response was set at 0.8 Hz–250 Hz Vibration Motor ERMS as show in Fig. 5.

Figure 6 shows the location of two sensors and vibration motors inside the pillow. Vibration motor ERMS 1 is located under the neck in order to detect vibration caused by snoring sound [4]. Sound sensor module are located in below position of the pillow to detect the sounds [9, 11].

According to Fig. 7, the researcher studied the obtained data for research development. The data was collected from the interviews with physiotherapists, nurses, and expert physicians for the design of the portable snoring sensor pillow. The operations of the equipment began with the vibration of the motor when snoring sounds were over the critical level as determined so that users were warned to change their sleeping postures (positions) and stop snoring [5]. The researcher designed and developed a set of equipment for the snoring sensor pillow. It consisted of a sound sensor module, a vibrating motor, and Arduino control panel. The procedures of the design together with the development were imparted as follows:

(1) Designed an electronic circuit to control and monitor the operation of the motor. The command was written to Arduino control panel in C programming language. Then, the operation was tested.
(2) Connected the sound sensor module circuit to Arduino control panel. Wrote the command so that the sound sensor module could adjust values of voices. Next, tested the operation. If snoring sounds were detected, the motor would vibrate to activate snoring sleepers and to stop their snoring [8].
(3) The last step was to insert the set of the equipment inside the designed pillow.

Fig. 3. Sound sensor module

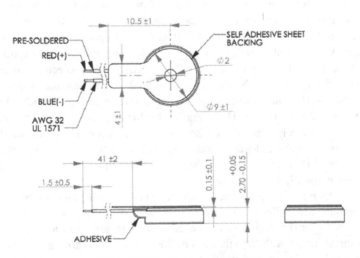

Fig. 4. Vibration motor ERMS

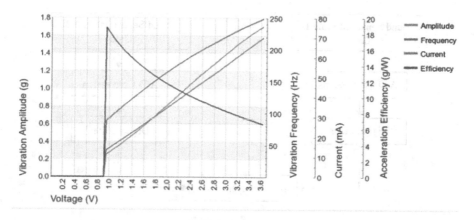

Fig. 5. Typical frequency response of vibration motor ERMS

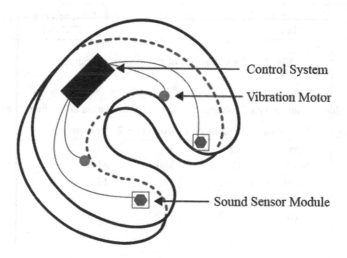

Fig. 6. Location of the vibration motor ERMS and sound sensor module inside the pillow

4.2 Data Acquisition

There were 3 samples who met the selection criteria of the research and treatment follow up. All of then were males and were tested by employing the snoring sensor pillow.

At the first stage, three subjects (males) summarized in Table 2. Data were participated in the laboratory environment to adjust the threshold value in order to separate the snoring only from the subject and ratio of snoring time [10]. Table 3 shows environment of stored data. Since it is known that snoring sounds measurements from each subject were recorded by the data acquisition system, biopac [11].

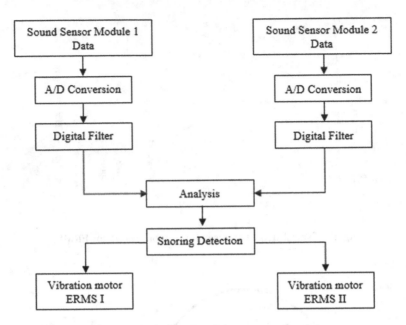

Fig. 7. Block diagram of the system operation

Table 2. Summary of volunteer's physical information at the first stage

Parameters	Volunteer 1	Volunteer 2	Volunteer 3
Age	27	30	34
Sex	Female	Male	Male
Height (cm)	162	175	172
Weight (Kg)	48	76	72
BMI	18.29	24.82	24.34

Table 3. Environment of stored data

Noise	Definitions
Case 1	Snoring only from the subject
Case 2	Snoring with ambient music

4.3 Data Analysis Algorithm

The operation of the pillow was tested in accordance with the specified functions. To illustrate, when the microphone module could detect snoring sounds, it would command the motor to vibrate for alerting the snoring sleepers to be conscious or to change sleeping postures and stop snoring [4]. The command was written in C programming language for controlling the vibration of the motor as show in Fig. 8.

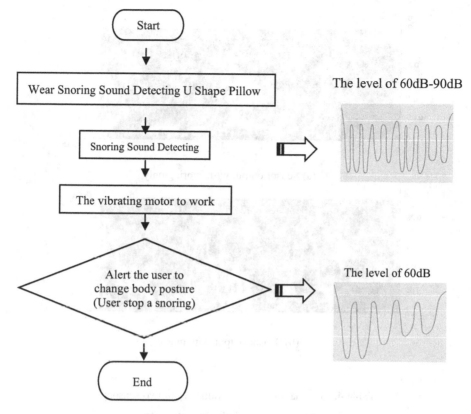

Fig. 8. Flowchart of data analysis

5 Experimental Results

The research purposed to performance is compared result by Snoring only from the subject and snoring with ambient music. The experimental results are as follows:

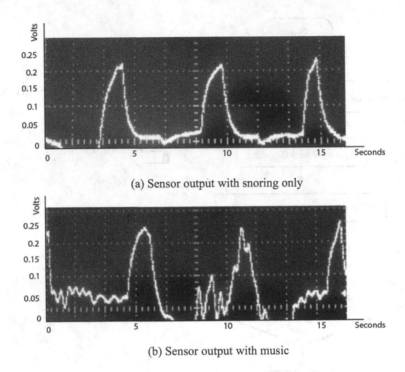

(a) Sensor output with snoring only

(b) Sensor output with music

Table 4. Summary of snoring values from two sensors

		Sound sensor	Ratio (%)	Remarks
a	1	0.195	92.27	Snoring
	2	0.196	92.87	Snoring
	3	0.213	95.54	Snoring
b	1	0.035	−12.32	Music
	2	0.219	92.54	Snoring
	3	0.081	−102.35	Music
	4	0.218	92.48	Snoring
	5	0.032	−12.10	Music

From the experiments that have tested for 2 level, we found the ratio of the snoring only from the subject and snoring with ambient music performances are described as follow;

As can be seen from the Table 4, the positive values of ratio are from 92.27% up to 95.54%. Ratios with negative values mean that the decision of noise [11]. Therefore, the ratio of higher than 70% is sufficient as a threshold for consideration snoring sounds from ambient noises.

6 Conclusion

In this paper, after the researcher had evaluated the efficiency of the pillow's performance, it was discovered that the distance between the microphone module and snoring sounds that could make the motor vibrate at its best was 10 cm. And when the pillow was actually applied to the 3 samples, it was unveiled that the results of the test depended on the loudness of snoring sounds (dB). Moreover, the longer the distances between snoring sounds and the sound sensor module were, the more difficult to detect snoring sounds. The vibration of the motor would also delay or even not operate at all. Apart from the aforesaid factors, it hinged on ages, genders, or sleeping postures of the samples as well.

References

1. Satoshi, I., Satoshi, U., Nakamura, Y., Motegi, M., Ogawa, K., Shimokura, K.: A basic study of a pillow-shaped haptic device using a pneumatic actuator. In: IEEE International Symposium on Mechatronics and its Applications, Amman, 27–29 May 2008
2. Azarbarzin, A., Moussavi, A.: Snoring sounds variability as a signature of obstructive sleep apnea. Med. Eng. Phys. **35**(4), 479–485 (2013)
3. Shumit, S., Mahsa, T., Zahra, M., Azadeh, Y.: Effects of changing in the neck circumference during sleep on snoring sound characteristics. In: International Conference of the IEEE Engineering in Medicine and Biology Society (EMBC), pp. 2235–2238 (2015)
4. Jin, Z., Qian, Z., Wang, Y., Qiu, C.: A real-time auto-adjustable smart pillow system for sleep apnea detection and treatment. In: IPSN, pp. 179–190 (2013)
5. Wei, R., Li, X., Kim, H.S., Im, J.J., Kim, H.J.: A development of pillow for detection and restraining of snoring. In: IEEE International Conference on Biomedical Engineering and Informatics, pp. 1381–1385 (2010)
6. Rajendra, S., Shashikant, D.: Mobile operated anti-snoring pillow. In: IEEE International Conference on Environmental and Computer Science, pp. 441–444 (2009)
7. Hyung, G., Dong, W.: Intelligent pillow type wireless charger for fully implantable middle ear hearing device with a function of electromagnetic emission reduction. In: IEEE International Symposium on Intelligent Information Technology Application, pp. 835–838 (2008)
8. Zhu, X., Chen, W.: Automatic home care system for monitoring HR/RR during sleep. In: International IEEE EMBS Conference Vancouver, pp. 522–525, 20–24 August 2008
9. Kongsak, R., Nuntigar S.: Short-term outcomes of transoral radiofrequency somnoplasty treatment for snoring (2009). Shortterm outcomes of Transoral radiofrequency somnoplasty-treatment for snoring. http://www.rcot.org/download/
10. Azarbarzin, A., Zahra, M.: A comparison between recording sites of snoring sounds in relation to upper airway obstruction. In: International Conference of the IEEE EMBS, pp. 4246–4249 (2012)
11. Jee, D.K., Kim, H.S., Wei, R., Im, J.J.: A study for the development of PVDF vibration sensor and establishment of noise removal algorithm for snoring detection pillow. In: IEEE International Conference on Environmental Engineering and Informatics, pp. 1031–1035 (2011)
12. Boomidevi, R., Pandiyan, R., Rajasekaran, N.: Monitoring cardio respiratory and gesture recognition system using sensor pillow system. Int. Res. J. Eng. Technol. (IRJET) **2**(8), 83–87 (2015)

The Electroencephalography Signals Using Artificial Neural Network for Monitoring Fatigue System

Worawut Yimyam[1(✉)] and Mahasak Ketcham[2]

[1] Department of Computer Business, Phetchaburi Rajabhat University,
Phetchaburi, Thailand
`worawut_yimyam@hotmail.com`
[2] Department of Information Technology Management, King Mongkut's
University of Technology North Bangkok, Bangkok, Thailand
`mahasak.k@it.kmutnb.ac.th`

Abstract. This paper proposes the development of an algorithm used for monitoring fatigue of the soldiers while they perform their duty. Electroencephalography (EEG) signals are analyzed by an Artificial Neural Networks (ANN) technique and compared with other techniques. The experimental results show that the ANN provides more accurate results than Bayesnet, Support Vector Machines (SMO), and Naïve Bayes techniques. The result of the ANN technique provides the accuracy, recall, and precision values at 83.77, 0.838, and 0.838, respectively.

Keywords: Fatigue · Electroencephalography · Artificial Neural Networks

1 Introduction

Nowadays, there are many important duties of solders in the mission of the military to perform against the law breaking such as the maintenance of the independence, sovereignty, national security, border security, and forest patrol. There are also the offenses that affect the national security such as drug and weapon trade, and deforestation. Each mission need lots of times to achieve on it. From that, soldiers may feel fatigued because of their mission [1]. When fatigue occurs, it causes accidents during the mission and also reduce work performance of the solders. For example, there was an accident occurred to US military called M985 in truck drive. The investigation found that the accident happened because of the driver's fatigue which leads him to death because of the lack of sleeping [3]. This type of problem has affected to the wrong decision, communication errors, and risk assessment. All these consequences have systematic relationships [6].

Military's mission may face the risk any time. One mission needs many hours for working and uses a lot of military personnel. However, a number of soldiers are not enough in some military base [13, 14] Hence, these actions cause fatigue because the soldiers might work several tasks. For example, the mission of the military is to train the pilots to fly helicopter. In facts, the pilots should sleep appropriately because they

© Springer International Publishing AG 2017
M. Numao et al. (Eds.): PRICAI 2016 Workshops, LNAI 10004, pp. 160–169, 2017.
DOI: 10.1007/978-3-319-60675-0_14

have to work all day based on FFA rule. They have to wake up at 5.00 am, monitor the helicopter's engine before flying at 4 pm, take off at 5.30 pm, land at 10.30 pm, and store the helicopter at 12.30 am. Then, they move to another base to prepare helicopter at 2.30 am, re-check again at 6.00 am, until military mission is completed. As from their daily routine duties, it can be seen that the pilot's problem is the lack of sleep. Thus, the system is developed to monitor fatigue from the eyes [10, 11]. There are many researches working in this area. However, the measurement of fatigue is not certainly accurate and effective.

Saeid Fazli et al. [8] proposed eyes tracking in order to monitor fatigue by using image processing technique to find the eyes position and check whether the eyes open or close. The results showed some errors because the experiments of the closed and opened eyes depend on an individual eye. Ye sun et al. [16] proposed the fatigue monitoring system to detect the physiological sign consisting of eyes and heart. Moreover, sensor technology has been used, but it has a problem in terms of the limitation on transmission distance. Edward et al. [4] proposed the evaluation technologies that can help to monitor fatigue. Researchers found that the eyes monitoring technology has high reliability in case of indicating the fatigue of the body. YenWei Chen and Kenji Kubo [17] proposed the development of face detection and eye movement via webcam by using Gabor filter technique. Filter technique works with color filter data. For face detection, the system monitors a geometric shape from face structure and uses Gabor filter to shift the image. The system runs continuously and displays the results of face detection via a monitor. Sung-Uk Jung and Jang-Hee Yo [9] proposed the method to increase the quality of eye detection. Researchers cut an irrelevant distraction by SQI method. The three-dimensional image conversion helps detect the eye position. Then, AdaBoost was used in identifying the eye position more precisely.

This paper proposes the development of an algorithm used for monitoring fatigue of the soldiers while they perform their duty. Electroencephalography (EEG) signals are analyzed by an Artificial Neural Networks (ANN) technique and compared with other techniques. Preliminary

1.1 Fatigue

In case of fatigue, sleepiness, and loss of concentration in perspective of Intelligent Transportation System: ITS, it was found that it has the same meaning of science, in which there is no criteria for measure of fatigue precisely [10, 11]. Fatigue can be measured by nerve, muscle, body temperature, eye movement, respiratory rate, heart rate and brain function [2, 5, 11, 12]. In addition, the brain function provides better results for analyzing the fatigue and sleepiness. According to the Psychomotor Vigilance Task (PVT) research, it has mentioned that the visual stimulation and visual responsiveness are the main factors of brain monitoring [7].

1.2   EEG Monitoring

Human brain whether it has been sleeping or awaking, it has different frequencies which can observe through EEG monitoring. The frequency of signal occurs when there is a change of the electrical stimulation. The analysis of EEG signal has been developed by many researchers [18–20]. EEG monitoring system has been utilized by Digital Signal Processing Units due to its frequency range which can show the behavior of patients and the signal measurement [21]. The types of EEG signal are shown in Table 1.

Table 1. Types of EEG Signal [24]

Types of EEG signal	Frequency spectrum (Hz)	Amplitude (μV)	Significance
Delta	0.1–0.3	100–200	Deepest, dreamless sleep, unconscious state, cognitive tasks by frontal lobe
Theta	4.0–7.5	<30	REM sleep, dreaming, physiological at the age of 1–6, cognitive task by frontal lobe (Fourier analysis), intuition, creativity
Alpha	8.0–12.0	30–50	The "basic" wave of the brain occurred when stimulating high frequency (alpha block). Relaxed but not sleepy state
Beta	13.0–30.0	<20	Sensory and emotional influences, harmonic, wide awake, exciting, conscious states
Gamma	30.0–50.0	<10	High mental activity

2   System Design

This research focuses on the fatigue monitoring system of the military mission. The system is implemented by the EEG sensor in which it can send signal to smartphone via Buletooth and the frequency radio wave, and can collect EEG data via smartphone program. The system is able to analyze the soldier's fatigue conditions and sends the alert to the admin. The Fig. 1 shows the overview of system.

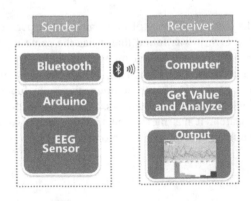

Fig. 1. Overview of system

2.1　EEG Sensor

This part is the analysis of the EEG signal received from EEG sensor. The fatigue from researcher's EEG signal is tested as shown in Fig. 2.

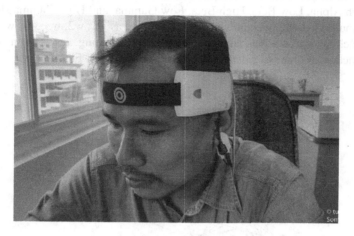

Fig. 2. MindFlex headset

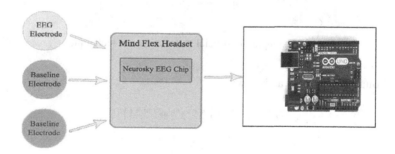

Fig. 3. Neurosky EEG chip

2.2　Communication Link Arduino

EEG monitoring data is received from MindFlex Headset device. Arduino board is also used to convert value received from sensor to different types of signals which can be divided into 8 ranges: Delta 1–3 Hz, Theta 4–7 Hz, Low Alpha 8–9 Hz, High Beta 18–20 Hz, Low Gamma 31–40 Hz, and High Gamma 41–50 Hz. These frequency waves are transformed in ASCII coding (Fig. 3).

2.3　Data Reception

In data reception, the system connects to the Bluetooth module in order to send signal to computer for analyzing the fatigue in the next step.

2.4   Receive Value and Analyze with ANN

Artificial Neural Network technique (ANN) is used to analyze data receiving from the EEG signals. The processing applies with neural network of the human brain. The EEG signal is an input of the ANN technique. The EEG data composes of Delta, Theta, Low Alpha, High Alpha, Low Beta, High Beta, Low Gamma, and High Gamma signals. All inputs are multiplied with weight which is represented as w1, w2, w3, w4, w5, w6, w7, and w8. Each neuron is a bias adjustment with the weighting. It has been sent to the transfer function in order to calculate the result as shown in Fig. 4.

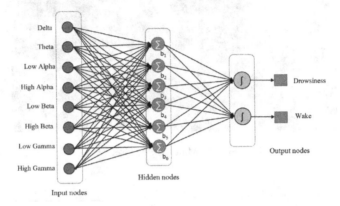

Fig. 4. Example of artificial neural network technique

The Equation is shown as below:

$$a^m = f^{m+1}\left(w^{m+1}x^m + b^{m+1}\right) \tag{1}$$

Where
a^m means Output Node
f^m means Transfer Function
$w^m = 0.2$
$b^m = 0.1$
x^m means Delta, Theta, Low Alpha, High Alpha, Low Beta, High Beta, Low Gamma, High Gamma.

3   Experimental Results

The dataset collected from user's EEG signals are measured using EEG sensor. The system requires the data record from user's EEG signals for 20 times, 10 times for falling asleep and 10 times for waking up. Each record contains five minutes. The dataset is experimented for finding the fatigue symptom. As a result, the signals can be divided from each record into 8 signal waves as shown in Fig. 5.

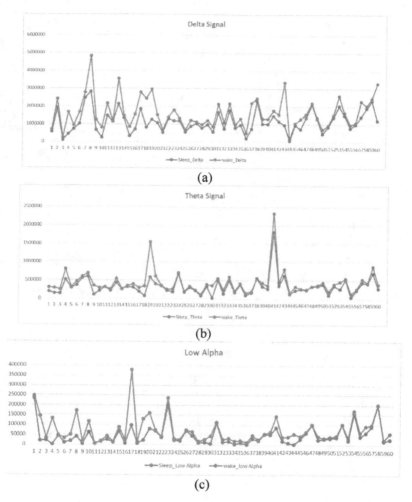

(a)

(b)

(c)

Fig. 5. (a) The graph of signals including (a) delta signal (b) theta signal (c) low alpha signal (d) high alpha signal (e) low beta signal (f) high beta signal (g) low gamma signal (h) high gamma signal

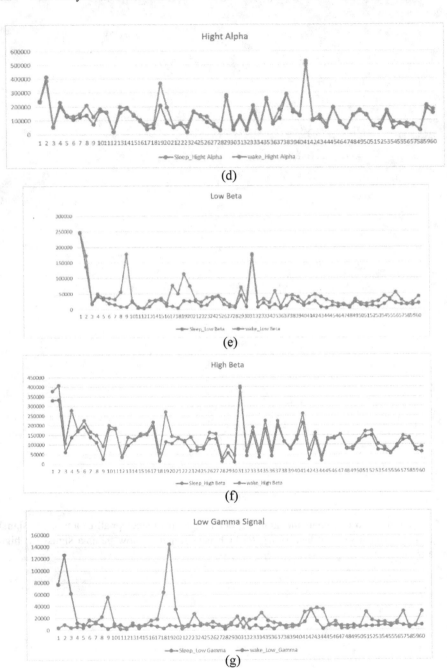

Fig. 5. (continued)

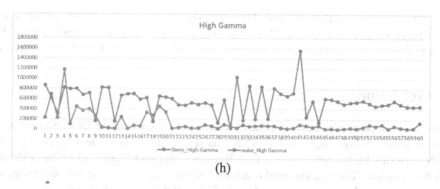

(h)

Fig. 5. (continued)

The performance of classification is conducted by the ANN technique. It considers the accuracy of precision and recall as shown Eqs. (2), (3), and (4). Table 2 shows the experimental results of the performance of predicted class. Table 3 shows the comparison of the performance of classification.

$$Precision(p) = \frac{TP}{TP + FP} \tag{2}$$

$$Recall(r) = \frac{TP}{TP + FN} \tag{3}$$

$$Accuracy(A) = \frac{TP + TN}{TP + TN + FP + FN} \tag{4}$$

TP = True Positive, FP = False Positive, FT = True Negative and FN = False Negative.

Table 2. The result of data experimental classification

Actual class		Predicted class	
		Drowsiness	Wake
Actual class	Drowsiness	3546	595
	Wake	667	2993

Table 3. The comparison of the performance of classification

Model	10-fold cross validation		
	Accuracy	Recall	Precision
ANN	83.77	0.838	0.838
Bayesnet	77.07	0.771	0.777
SMO	75.40	0.754	0.759
Naive Bayes	62.10	0.621	0.712

From the experiment, it was found that the ANN technique has higher performance than Bayesnet, Support Vector Machines (SMO), and Naïve Bayes techniques. the values of the accuracy, recall, and precision are 83.77, 0.838, and 0.838%, respectively.

4   Conclusion

Researcher proposes the development of algorithm for monitoring fatigue in military mission based on brain signals. The ANN was applied to analyze the data. As a result, ANN performs higher performance than Bayesnet, Support Vector Machines (SMO), and Naïve Bayes. The experimental result of ANN technique showed the percentage of its accuracy, recall, and precision values at 83.77, 0.838, and 0.838, respectively.

References

1. Sicard, B.: Risk propensity assessment in military special operations. Mil. Med. **166**(10), 871 (2001)
2. Lin, C.T., Ko, L.W., Chung, I.F., Huang, T.Y., Chen, Y.C., Jung, T.P., Liang, S.F.: Adaptive EEG-based alertness estimation system by using ICA-based fuzzy neural networks. IEEE Trans. Circuits Syst. **53**(11), 2469–2476 (2006)
3. Department of the Army: Leaders' Manual for Combat Stress Control, FM22-51, Washington DC, September 1994
4. Edwards, D.J., Sirois, B., Dawson, T., Aguirre, A., et al.: Evaluation of fatigue management technologies using weighted feature matrix method. In: Proceedings of the Fourth International Driving Symposium on Human Factors in Driver Assessment, Training and Vehicle Design, Stevenson, Washington (2007)
5. Cai, H., Lin, Y.: An experiment to non-intrusively collect physiological parameters towards driver state detection. In: Proceedings of SAE 2007 World Congress, No. 2007-01-0403. SAE Technical Paper (2007)
6. How, J.M., Foo, S.C., Low, E., Wong, T.M., Vijayan, A., Siew, M.G., Kanapathy, R.: Effects of sleep deprivation on performance of Naval seamen: I. Total sleep deprivation on performance. Ann. Acad. Med. **23**(5), 669–675 (1994). Singapore
7. Rau, P.S.: Drowsy driver detection and warning system for commercial vehicle drivers: field operational test design, data analyses, and progress. In: National Highway Traffic Safety Administration, pp. 05–0192 (2005)
8. Fazli, S., Esfehani, P.: Tracking eye state for fatigue detection. In: International Conference on Advances in Computer and Electrical Engineering (ICACEE 2012), pp. 17–20 (2012)
9. Jung, S.U., Yoo, J.H.: Robust eye detection using self quotient image. In: Intelligent Signal Processing and Communications (ISPACS 2006), Japan, pp. 263–266. IEEE (2006)
10. Brandt, T., Stemmer, R., Rakotonirainy, A.: Affordable visual driver monitoring system for fatigue and monotony. In: IEEE International Conference on Systems, Man and Cybernetics, vol. 7, pp. 6451–6456. IEEE (2004)
11. Von Jan, T., Karnahl, T., Seifert, K., Hilgenstock, J., Zobel, R.: Don't sleep and drive–VW's fatigue detection technology. In: Proceedings of 19th International Conference on Enhanced Safety of Vehicles, Washington, DC (2005)

12. Nakagawa, T., Kawachi, T., Arimitsu, S., Kanno, M., Sasaki, K., Hosaka, H.: Drowsiness detection using spectrum analysis of eye movement and effective stimuli to keep driver awake. DENSO Tech. Rev. **12**(1), 113–118 (2006)
13. US Army Safety Center: Sustaining performance in combat. Flight Fax **5**(31), 9–11 (2003)
14. US Army Safety Center: Fatigue. Countermeasure **3**(23), 4–5 (2002)
15. Raudonis, V., Simutis, R., Narvydas, G.: Discrete eye tracking for medical applications. In: 2nd International Symposium on Applied Sciences in Biomedical and Communication Technologies (ISABEL 2009), pp. 1–6. IEEE (2009)
16. Sun, Y., Yu, X., Berilla, J.: An innovative non-invasive ECG sensor and comparison study with clinic system. In: 39th Annual Northeast Bioengineering Conference (NEBEC), pp. 163–164. IEEE (2013)
17. Chen, Y.W., Kubo, K.: A robust eye detection and tracking technique using gabor filters. In: Third International Conference on Intelligent Information Hiding and Multimedia Signal Processing (IIHMSP 2007), vol. 1, pp. 109–112. IEEE (2007)
18. Adeli, H., Ghosh-Dastidar, S., Dadmehr, N.: A spatio-temporal wavelet-chaos methodology for EEG-based diagnosis of Alzheimer's disease. Neurosci. Lett. **444**(2), 190–194 (2008)
19. Dauwels, J., Vialatte, F., Musha, T.: A comparative study of synchrony measures for the early diagnosis of diagnosis of Alzheimer's disease based on EEG. Neuroimage **49**(1), 668–693 (2010)
20. Morison, G., Tieges, Z., Kilborn, K.: Multiscale permutation entropy analysis of the EEG in early stage Alzheimer's patients. In: Proceedings of the IEEE Engineering in Medicine and Biology Society, CA (2012)
21. Kantona, J., Farkas, T., Dukan, P., Kovari, A.: Evaluation of the Neurosky MindFlex EEG headset brain waves data. In: IEEE 12th International Symposium on Applied Machine Intelligence and Informatics (2014)
22. NeuroSky Inc.: The brain wave signal (EEG). NeuroSky Inc. (2009)
23. Onchira, O., Mawaporn, W.: Indoor localization of a wireless ad hoc sensor networks. In: The 9th National Conference on Computing and Information Technology (2013)
24. Buzsaki, G.: Rhythms of the Brain. Oxford University Press, Oxford (2006)

Arrival Time Prediction and Train Tracking Analysis

Somkiat Kosolsombat[(✉)] and Wasit Limprasert

Department of Computer Science, Faculty of Science and Technology,
Thammasat University, Pathumthani, Thailand
{somkiat.k,wasit_l}@sci.tu.ac.th

Abstract. Rail transportation is a convenient and safe in many countries. However, Rail transportation in some countries has significant long delays. Arrival time prediction and rescheduling the time table are partial solutions to tackle the delay problem. In this paper, the relationship between measurable properties and the delay time are studied in order to develop an arrival time prediction. The result of this experiment has three parts. The relationship between properties and arrival late are then visualized and discussed. Some properties from the acquired database show that week, day and station, are important features and impact on the delay. Various regression methods are compared in our experiment and the result shows that best RMSE is ± 3.863 min by applying Random Forest Regression on train tracking dataset.

Keywords: Arrival time prediction · Train tracking analysis · Arrival regression

1 Introduction

According to the data acquired from Sate Railway of Thailand (SRT) website and train tracking database system. The data has been collected for 1 year in 2015 consisting over 975,386 records. The mean different between schedule time and actual arrive time is about 18 ± 16 min. This result occurred in huge economic impact and transportation delay.

One of the cause the arrival delay is that many trains running on the same track which has possibility to have more than one train in the same location. Mostly the railway agents have to solve the route conflict manually in real-time. From our preliminary analysis, we found that route conflict is the main reason that causes arrival delay about 50,000 incidents annually.

Many researches attempted to find solutions to reduce the arrival delay. In 2014 a study [1] used heuristic approach to find suitable solution for scheduling in order to prevent deadlock in a single track railway problem. The solution attempts to solve the route conflict and find the near-optimal travel strategies. Similar approach also considered in [2], which proposed a train scheduling system for double-track railway. The system is able to predict a route conflict and the system is also able to reschedule to minimize the conflict using a stochastic graph.

© Springer International Publishing AG 2017
M. Numao et al. (Eds.): PRICAI 2016 Workshops, LNAI 10004, pp. 170–177, 2017.
DOI: 10.1007/978-3-319-60675-0_15

One of the most important elements in train scheduling system is the ability to predict the arrival time of all trains arriving all stations. There are many studies attempting to find the best machine learning method to predict the arrival time. For example, In 2014 a study [3] developed train arrival time prediction model by comparing k-NN and moving average of time series data. In this experiment used arrival records collected from three different routes (#75, #201 and #407). The result shows there is no significant different between k-NN and the moving average technique. A study in [4] compares performance between SVR and ANN implemented on MATLAB by using train arrival delay records and other related information from Surbian Railways Network consists 727 routes of the passenger trains. The result of SVR is better than ANN in this particular experiment. In [5], the data retrieved from Iranian Railways between 2005 and 2009 is used evaluating the delay prediction model using ANN, decision tree and logistic regression. The result suggested that ANN is outperform other two methods. From the related studies, SVR and ANN are likely to be the suitable regression method for this type of problem.

In this paper, we are going to compare three regression methods in order to find the possible good candidate for a train arrival time prediction system.

2 Our Data Analysis Method

In this paper, the dataset in our experiment is acquired from the official web site, expert interview and train tracking database system of State Railway of Thailand (SRT). The data has been constantly collected for 1 year of 2015 consisting around 975,386 records and a summarized histogram of arrival late is shown in Fig. 1. The original dataset consists many tables. Some table are dropped out because it is unrelated or consisting of too many missing data. For example, in *train_tracking* table, *arrive_note* field and *leave_note* field are all empty.

There are three tables; *train_running*, *train_tracking* and *time_table*. The *train_-tracking* table and *train_running* table are merged by *train_running* id. After merging, the number of record is 975,386 records. After that, we merge it with *time_table* by *time_table* id. The number of record after merging is 686,445 records. Then the merged table are validated and dropped all records containing Null or NAN or duplicate values. All fields containing date-time are converted to numeric data type. The *started_on* id is converted to *week* and *day*, to represent index of week in year and index of day in week. Finally, the number of record after cleaning is 323,543 records. The result after this pre-processing is shown in Table 1, which contains 11 fields. The arrive time is set to be output of the regression called y and the remaining columns are formed set of vector x.

We divide our study into three experiments. In the first experiment, the data is analyzed for basic visualize to represent important characteristics. In the second experiment the ExtraTrees classifier from Scikit-learn [6] is applied to extract key features that have significant impact on the arrival late. In the final experiment, three regression techniques are compared used the acquired for the evaluation.

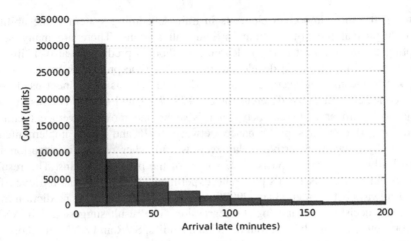

Fig. 1. The histogram of arrival late (minutes)

3 The Experiment and Result

3.1 Preliminary Visualization

In this experiment, we defined two properties to express basics characteristic of the train arrival data.

Delay magnitude is the function of arrive late respected to value of particular feature. The delay magnitude is illustrated by a cumulative density function (CDF), where the 0.9 line expresses 90% of the trains arrive to the station by the time.

Activity function is the function to represent the relation between the number activities respected to value of particular feature. The activity function is computed by calculating a histogram of the number of arrival, where the bin is interval on the particular feature.

Table 1. List of feature after pre-processing

Symbol	Feature	Original data type	After pre-processing
x1	*default_leavetime*	Date-time	Integer
x2	*default_arrivetime*	Date-time	Integer
x3	*leave_time*	Date-time	Integer
x4	*arrive_time*	Date-time	Integer
x5	*leave_cause*	Integer	Integer
x6	*arrive_cause*	Integer	Integer
x7	*train*	Integer	Integer
x8	*station*	Integer	Integer
x9	*day*	Integer	Integer
x10	*week*	Integer	Integer
y	*arrive_late*	Integer	Float

From Figs. 2, 3, 4 and 5, the figures show multiple-axis graph of the delay magnitude and activity function. The delay magnitude has a unit in minutes as represent in the left vertical axis. the activity function shows the number of arrival and the unit is in right vertical axis.

Figure 2 shows the delay magnitude and the activity function varying respect to week. The horizontal axis represent the index of week in a year. The dashed line indicates a number of arrival per week. The red solid line represents the delay magnitude with 0.9 confidence, 90% of trains arrive late less than the graph. Between 8th to 26th week, we found there is large magnitude of delay, while the activity is low, which in the same period of a long holiday. According to our interviews with some officers in SRT, during this period SRT needs to increase the number of carriage of the train. This increase loading time and reduce maximum speed of the trains.

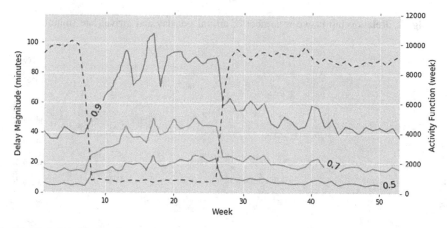

Fig. 2. Relationship between number of arrival per week and arrival late (minutes) (Color figure online)

Figure 3 shows the relationship of delay magnitude and activity function with the horizontal axis is index of day in week, where 0 is Monday and 6 is Sunday. The dashed line indicates a number of arrival per day. From the data, Monday Tuesday and Friday are high activity day. The magnitude of delay is almost the same during the week only about 10 min different between Friday and Saturday. In summary, more than 90% of arrival has delay time least than 50 min.

In Figs. 4 and 5 the horizontal axes are the default arrive time and the actual arrive time, respectively. The graphs in both figures have similar patterns. The activity function in both graphs are peak higher 18,000 arrival per hour at 6 am and 4 pm and the activity is dropped to 12,000 arrival per hour around midday. The delay magnitude is an increasing function with time between 6 am and 4 am of the next day. There are an abnormal characteristic around the midday, where the activity is off peak but the delay magnitude reaching the peak. This requires further investigation.

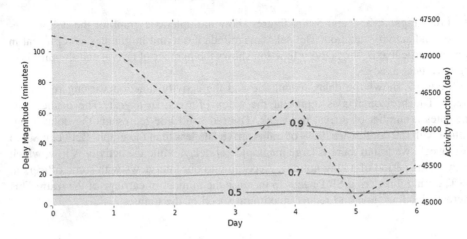

Fig. 3. Relationship between number of arrival per day and arrival late (minutes)

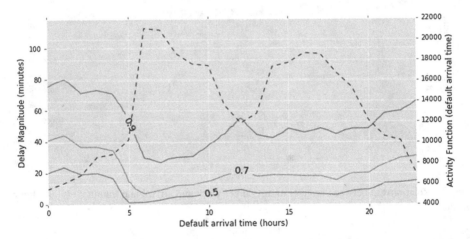

Fig. 4. Relationship between number of arrival per default arrival time and arrival late (minutes)

3.2 Feature Importance

The feature importance score (FIS) is average impurity reduction for all partition in the decision trees [6, 7]. The feature receiving high FIS is the most frequently occurring and producing large impurity reduction.

After pre-processing data, All features (X) is transformed to matrix. Similarly, the field *arrive_late* is converted to a vector (y). The ExtraTrees classifier [6] is applied to extract the feature importance score (FIS). The result ranking is sorted from most influent features, which is on the top to the least as shown in Table 2 *Week* is the most impact feature for predicting arrival time about receiving a FIS of 0.273. *Day, station* and *train* features have FIS at the same figure.

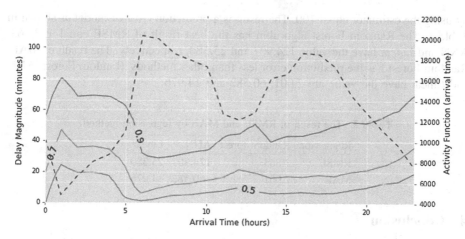

Fig. 5. Relationship between number of arrival per arrival time and arrival late (minutes)

Table 2. The most influent features sorted by the number of impacts

Ranking	Feature	FIS
1	*week*	0.273
2	*day*	0.117
3	*station*	0.113
4	*train_no*	0.107
5	*arrive_time*	0.072
6	*leave_time*	0.054
7	*arrive_cause*	0.049
8	*default_arrivetime*	0.041
9	*leave_cause*	0.041
10	*default_leavetime*	0.041

3.3 Regression Comparison

For most of regression studies, Root Mean Square Error (RMSE) (1) or Mean Absolute Error (MAE) (2) are commonly used to represent the error of model to fit the training data.

$$RMSE = \sqrt{\sum_{i=1}^{N} \frac{(y_{pi} - y_i)^2}{N}} \tag{1}$$

$$MAE = \frac{1}{N}\sum_{i=1}^{N} |y_{pi} - y_i| \tag{2}$$

After pre-processing data, all features (y_i) is transformed to matrix. Similarly, the field *arrive_late* is converted to a vector (y_{pi}). The ExtraTrees classifier [6] from Scikit-learn is applied to fit the training data and to find out the RMSE and MAE value by using the estimator equal 100. The result is a prediction error of model as shown in Table 3. The Random Forest algorithm has the best result of RMSE equal ± 3.863. ANN and Linear have the value 124.907 and 25.380, respectively. The result of MAE, Random Forest has the prediction error less than other methods. Random Forest, ANN and Linear have the value are 2.001, 60.582 and 14.976.

Table 3. The result of RMSE and MAE by regression method

Regression	Random forest	ANN	Linear
RMSE	3.863	124.907	25.380
MAE	2.001	60.582	14.976

4 Conclusion

This paper examines the arrival time prediction and train tracking analysis. All data retrieved from the official web site, expert interview and train tracking database of State Railway of Thailand (SRT) for 1 year of 2015 consisting 975,386 records. The relationship between many properties and the delay magnitude (minutes) are studied. The comparison of three regression methods; Random Forest Regression, ANN, Linear regression, are also studied. The mean RMSE are 3.863, 124.907 and 25.380, respectively. We also found the accuracy of the classifier are affected by the importance feature. The most important feature is week receiving importance score of 0.273. Whereas, day, station *and* train_no receives no different importance score at 0.11. In future work, further analysis on arrival time and leave time will be examined in order to improve the accuracy of the current time table.

5 Discussion

This experiment retrieves data from three tables; *train_running*, *train_tracking* and *time_table*. The total after cleaning data is 323,543 records from all 975,386 records. It has many lost of data. Data is not complete or containing Null or NAN or duplicating value. It may have caused by data log recording, missing data, converting data, assigning the wrong data type, and so on. Then, data store or data pre-processing method are important to do experiment.

The result of regression model RMSE equal ± 3.863 min to indicate the value of prediction error for arrival time. The optimize this value may be consisting of the corrective data, rescheduling the time table, improving the recorded data, and adjusting the relationship between the actual time and default time table.

Acknowledgement. We would like to very thank Mr. Chokdee Suwanrat and his department of information technology and State Railway of Thailand for providing the data and consulting on a workflow of train operations and many definitions of the technical term.

References

1. Li, F., Sheu, J.-B., Gao, Z.-Y.: Deadlock analysis, prevention and train optimal travel mechanism in single-track railway system. Transp. Res. Part B Methodol. **68**, 385–414 (2014)
2. Kecman, P., Goverde, R.M.P.: Online data-driven adaptive prediction of train event times. IEEE Trans. Intell. Transp. Syst. **16**(1), 465–474 (2015)
3. Pongnumkul, S., Pechprasarn, T., Kunaseth, N., Chaipah, K.: Improving arrival time prediction of Thailand's passenger trains using historical travel times. In: 2014 11th International Joint Conference on Computer Science and Software Engineering (JCSSE), pp. 307–312 (2014)
4. Marković, N., Milinković, S., Tikhonov, K.S., Schonfeld, P.: Analyzing passenger train arrival delays with support vector regression. Transp. Res. Part C Emerg. Technol. **56**, 251–262 (2015)
5. Yaghini, M., Khoshraftar, M.M., Seyedabadi, M.: Railway passenger train delay prediction via neural network model. J. Adv. Transp. **47**(3), 355–368 (2013)
6. Pedregosa, F., Varoquaux, G., Gramfort, A., Michel, V., Thirion, B., Grisel, O., Blondel, M., Prettenhofer, P., Weiss, R., Dubourg, V., Vanderplas, J., Passos, A., Cournapeau, D., Brucher, M., Perrot, M., Duchesnay, É.: Scikit-learn: machine learning in Python, arXiv: 1201.0490 Cs (2012)
7. Breiman, L.: Random forests. Mach. Learn. **45**(1), 5–32 (2001)

Estimating PSD Characteristics of ECG in Comparison Between Normal and Supraventricular Subjects

Thaweesak Yingthawornsuk[✉], Siriphan Phetnuam, Saowaros Singkhal,
and Waraporn Pattarason

Media Technology, King Mongkut's University of Technology Thonburi, Bangkok, Thailand
thaweesak.yin@kmutt.ac.th, siriphan.jaa@gmail.com,
jeabsaowaros@gmail.com, krajib.pat@gmail.com

Abstract. The aims of project are to develop an arithmetic program that can detect irregularity in electrocardiogram (ECG) and classify between two groups of normal and supraventricular ECG waveforms by using Auto regressive (AR) estimators with various model orders starting from 3rd to 9th. All AR estimators are associated with the PSD of ECG waveforms collected from a group of 30 subjects at 200 Hz sampling frequency. The best classification scores found on the 5th-order AR model are 95.99% and 72.17% obtained from training and testing the C4_5 classifier with the fifth-order coefficients. By classifying the 7th-order AR coefficients with Linear Least Squared (LS) classifier the accurate scores of 86.43% and 80.85% were obtained from training and testing cases respectively. These performance accuracies show that the proposed method is highly effective in parameterizing and classifying PSD feature as quantitative measure that can characterize the ECG signals of normal and supraventricular cardiac conditions.

Keywords: ECG · PSD · AR · Supraventricular

1 Introduction

Cardiac Arrhythmia is a common type of heart disorder found in all aging persons and some with cardiac disorders. The cause of arrhythmia is various in each person such as the one who has a heart wall with very thickness or aging persons with cardiac disorder since they were born. The ECG signal is commonly used to represent on how our heart functions and reflects the healthiness condition of heart itself shown as a form of bioelectric signal. The shape of ECG waveform can feature a heart in terms of functionality, structural components or illness affection. This kind of signal can indicate changes in cardiac condition that mediate in its waveform.

Some of the most distressing types of heart malfunction occur not as a result of abnormal heart muscle but instead abnormal rhythm of the heart. Abnormality of any portion of the heart, including arterial and ventricle can sometimes causes a rapid rhythmic discharge of the impulse that spreads in all directions throughout a heart. This rapid heart rate, as determined from the time intervals between QRS complexes, is approximately 150 per minute instead of 72 per minute [4]. Supraventricular ECG is

© Springer International Publishing AG 2017
M. Numao et al. (Eds.): PRICAI 2016 Workshops, LNAI 10004, pp. 178–185, 2017.
DOI: 10.1007/978-3-319-60675-0_16

categorized in rapid tachycardia with heart rate above 100 per minute which is caused by electrical impulses originated above the heart's ventricles [6].

In this work, two different types of ECG signal comprised of normal ECG and Supraventricular ECG were comparatively studied to determine any significant difference in their power spectral distribution and other distinctions.

The following sections organized in paper are methodology, experimental results and discussion, and conclusion.

2   Methodology

The study procedure mainly consists of data acquisition, preprocessing signal, feature extraction including a model fitting and then classification. After all algorithms designed in each step correctly, the GUI for Matlab was designed and implemented to make it friendly for users. The main task of this study is to detect QRS complexes and the estimation of Power Spectral Density via AR modeling. The QRS complex detection is a major challenge. The Hamilton-Tompkins algorithm to detect complexes is divided into following steps [1, 4].

The pre-processor section performs linear and nonlinear filtering of the ECG signal and produces a set of periodic vectors. Next, the decision rule operates on the output

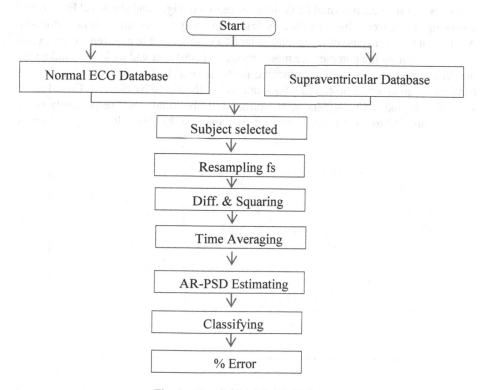

Fig. 1. Flowchart of the algorithm

from first section by classifying it as either a QRS complex or noise, and then save it for further steps. Filtering is used to attenuate noise. The Low and high filters are combined together to form a bandpass filter. The procedure of signal processing was followed by a differentiation, squaring and time averaging of the signal. Time averaging was done by adding the 32 most recent values from squared values and divided by 32. Figure 1 shows a workflow of PSD estimation and its comparison between two different databases. All ECG signals were down-resampled at 200 Hz which is adequate in acquisition of all information and frequency response components contained in signals. All procedure steps were repeated for individual recorded ECG signals in database until all thirty ECG file completed.

The signal databases used in this study consist of fifteen normal ECG signals and another fifteen supraventricular ECG signals. In main step of PSD estimation, the AR model based on Yule Walker's technique was applied to signals to determine the best fitted model coefficients that represent the signal in a form of spectral power distributed along a low frequency range [2, 3]. The coefficients belonging to the best fitted- order of AR model were used as a set of feature input to classifier in pairwise manner. The trials of possible fitting models to signal were performed on the model orders of AR.

3   Experimental Results and Discussion

In processing state each normal ECG signal as shown in Fig. 2 and abnormal ECG signal shown in Fig. 3. were observed visually first, then analyzed via our designed program. As one can notify, in case of supraventricular ECG the T wave in a current complex and P wave of a following complex are messy mixed to each other and look noisy alike with the dropped amplitude level as compared to the normal ECG. The S wave obviously deminishes and there is no reseting back interval to the signal baseline. In Figs. 4 and 5 show the original ECG signals from Normal and abnormal cases respectively being down resampled to have a new sampling frequency at 200 Hz, filtered by derivative

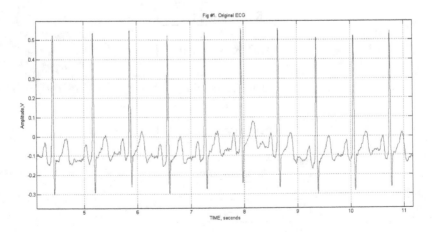

Fig. 2. Normal ECG waveform

filter, squared in its amplitude to have an absolute value in term of power and then time averaged with every 32 data-points window frame to have very evenly smoothed peaks. Results from processing two different groups of the categorized ECG signals via our program revealed the significant difference in term of quantitation of peaks found in state of time averaging. As one can see in case of abnormal ECG, it has a very less number of time averaged peaks as compared to normal one. In Fig. 6, the middle points of positive and negative slopes in individual time averaged peaks were detected automatically as indicated in red and blue markers shown in the same lower subplot.

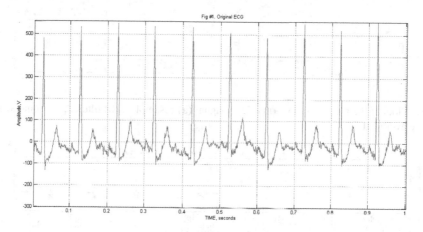

Fig. 3. Supraventricular ECG waveform

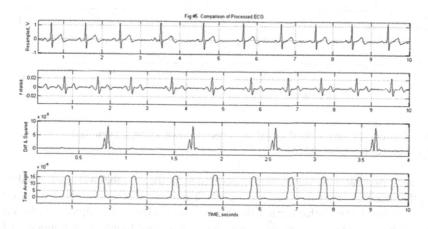

Fig. 4. Plots of the processed normal ECG signal (a) resampled, (b) filtered, (c) squared, and (d) time averaged

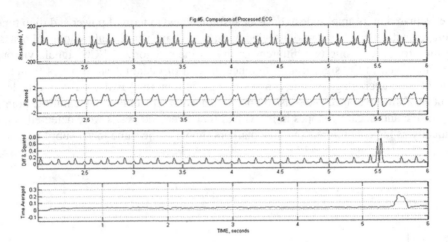

Fig. 5. Plots of the processed abnormal ECG signal (a) resampled, (b) filtered, (c) squared and (d) time averaged

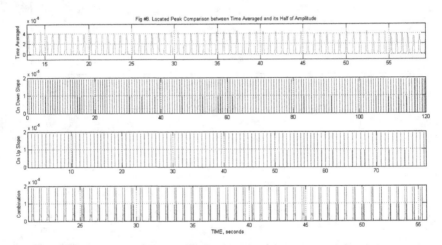

Fig. 6. Plots of (a) time averaged signal, (b) detected down-slope, (c) detected up-slope and (d) both markers (Color figure online)

Figure 7 illustrates the values of the AR- model coefficients that were estimated from fitting the models to the time averaged signals. The highest model order of AR used in signal modeling is the 12-th order and the Power Spectral Density of signal was therefore estimated based on the coefficients of AR model that were found in each model order. Next step is classification made in pairwise manner and the 20% of total number of AR coefficients was randomly selected and used to train the selected classifiers and the rest of coefficients used to evaluate the classification. The best classification scores found on the 5th-order AR model are 95.99% and 72.17% obtained from training and testing the C4_5 classifier with fifth-order coefficients. By classifying the 7th-order AR coefficients with Linear Least Squared (LS) classifier the accurate scores of 86.43% and 80.85% were obtained from training and testing cases respectively (Fig. 8).

```
arAll =

Columns 1 through 12

  1.0000   -2.5679    2.3331   -0.7637        0        0        0        0        0        0        0        0
  1.0000   -3.2473    4.4087   -3.0482   0.8896        0        0        0        0        0        0        0
  1.0000   -3.8967    6.6338   -6.2663   3.2600  -0.7300        0        0        0        0        0        0
  1.0000   -4.4697    9.1930  -11.1857   8.4678  -3.7890   0.7850        0        0        0        0        0
  1.0000   -5.0543   12.0147  -17.4916  16.7977 -10.6350   4.1136  -0.7447        0        0        0        0
  1.0000   -5.5473   14.7375  -24.5310  27.9163 -22.2129  12.0662  -4.0902   0.6619        0        0        0
  1.0000   -5.9393   17.1602  -31.6780  41.0733 -38.7481  26.5964 -12.8195   3.9476  -0.5923        0        0
  1.0000   -6.0915   18.1745  -34.9720  47.9073 -48.7045  37.1502 -20.9592   8.3570  -2.1184   0.2570        0
  1.0000   -6.0696   17.9939  -34.2594  46.1200 -45.5365  32.9969 -16.8739   5.3747  -0.5686  -0.2625   0.0853
  1.0000   -6.0902   18.0573  -34.1220  44.8212 -41.4889  25.0231  -5.8699   5.7703   7.7103  -4.6108   1.5520
       0        0        0        0        0        0        0        0        0        0        0        0
       0        0        0        0        0        0        0        0        0        0        0        0

Column 13

    0
    0
    0
    0
    0
    0
    0
    0
    0
-0.2417
    0
    0
```

Fig. 7. The estimated AR coefficients based on Yule Walker's technique

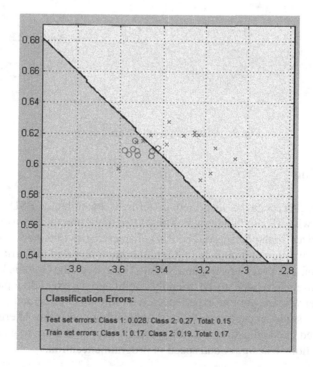

Fig. 8. Region of class discrimination and errors in classification

In order to simplify all complicated steps in executing all Matlab scripts to perform all steps in a workflow depicted in Fig. 1, the Graphic User Interface (GUI) was purposely designed to be much friendly and facilitated for user. Figure 9 presents GUI screens with plots of processed signals in result of executing each functional button that was designed to callback the Matlab scripts to perform tasks in selected function and then plot out the result of that state as shown in Fig. 9. User can make an observation on how original ECG signal is analyzed step by step clearly.

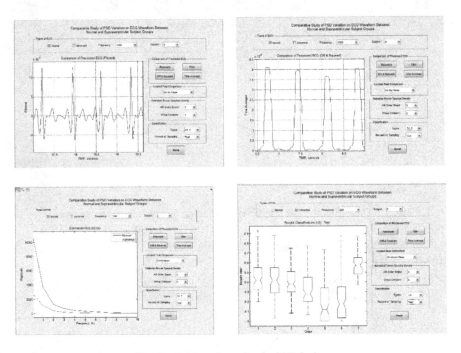

Fig. 9. Analyzed results on GUI design

4   Conclusion

Different significance can be identified in term of quantitative PSD estimated by modeling ECG with AR coefficients which best fitted to the processed waveform of ECG in each categorized signal database. The PSD characteristics obtained from estimation made on between different groups of ECG signals have revealed the certainly different frequency response at a very-low frequency range. Moreover, the Graphic User Interface (GUI) is purposely provided to be friendly facilitated in execution the processing program by user who may not be experienced with Matlab environment.

As shown in study on performance and computation cost, other alternative techniques are required in combination with the presently proposed work to gain more satisfactory expectation on accuracy and speed of program execution.

References

1. Patrick, S.H., Willis, J.T.: Quantitative investigation of QRS detection rules using the MIT/BH Arrhythmia database. IEEE Trans. Biomed. Eng. **BME-33**(12), 1157–1165 (1986)
2. Chusak, T., Thaweesak, Y.: Cardiac Arrhythmia classification using beat-by-beat autoregressive modeling. In: 3rd International Conference on Computer and Electrical Engineering (2010)
3. Junyou, H.: Study of Autoregressive (AR) spectrum estimation algorithm for vibration signals of industrial steam turbines. Int. J. Control Autom. **7**(8), 349–362 (2014)
4. Guyton, C.: Textbook of Medical Physiology. 8th edn. Harcourt College Pub. (1990)
5. Tompkins, W.J.: The Hamilton-Tompkins algorithm to detect complexes is divided into following steps. In: Tompkins, W.J. (ed.) Biomedical Digital Signal Processing. Prentice Hall, New Jersey (2000)
6. Obel, O.A., Camm, A.J.: Supraventricular tachycardia ECG diagnosis and anatomy. Eur. Heart J. **18**, 2–11 (1997)

Author Index

Printed in the United States
By Bookmasters

Printed in the United States
By Bookmasters